AIOps Checklists for Consulting & Projects

A Practical Guide to AIOps Transformation

Copyright © 2020 by Angel Berniz & Phee-Lip Sim

Disclaimer

The best of my efforts were put forth to validate the content provided in this book; the authors do not assume any responsibility for errors, omissions, or contrary interpretations of the subject matter contained within.

The information provided in this book is for educational purposes only. Authors do not accept any responsibilities for any liabilities or damages, real or perceived, resulting from the use of this information.

Table of contents

Preface

This book's purpose is to guide readers and expose them to tools needed to lead successful AIOps projects in many of the world's current organizations.

If you already know about me, you know that I love checklists as a powerful transformation tool.

We all know that checklist is a set of questions or list of statements that will help us keep track of all the actions that have to be performed in a work. The purpose of these checklists is to reduce failures, increase consistency and completeness in performing a specific task.

Now, you might wonder why make a book on checklists? The story goes something like this: one of my clients (a person in a CXO role) from a startup company, in a general discussion revealed a fact that their company had spent more than two thousand dollars in a year on buying checklists. Yet, another client said that they had memberships where the price tag is close to one thousand dollars every year to access checklists for their business technology projects.

These stories made me realize the importance of checklists for a company, and for that matter any company, team, or person. Also, this will mark my very first effort on defining checklists for AIOps projects. This book comes from an amalgamation of my twenty years of experience in the industry coupled with discussions with many other experts in the same field: business analysis, architecting, coding, testing, system integration, project management and transition.

Now you might assume that this book is only helpful for startups. However, I would like to emphasize that this book is not only useful for startups but also for strategists, business analysts, project managers, architects, developers, testers, and quality assurance professionals. I am confident that it will certainly help to set some good standards in their teams and perform their tasks effectively in line with industry's best practices.

If you are a person who wants to define a practical framework for your AIOps project success and engage your team like strategists, business analysts, architects, developers, testers, system integrators, project managers and so forth, then you are at the right place.

Acknowledgments

This book would not been possible without the help and support of so many friends of the AIOps community of practice.

My deepest gratitude goes to the team of contributors as they are a very special group of professionals who are true standouts in the AIOps community – for immersing themselves in this book project and helping me to bring it to fruition.

Also, a very special thank you to all the book reviewers who so generously contributed their time, energy, and expertise to make this book possible so that others might learn from their experience:

- Deeraj Bommaraju (general book review)
- Bhavani R (testing intro)
- Thomas Well (editor)

Foreword

We're in the digital era and despite all the certifications, degree programs (at undergraduate, graduate and even doctoral levels), projects still fail. Despite organizational focus on de-risking projects, projects still fail. Despite CxO's that should know better, organizations still handle initiatives terribly and actively undermine projects. The good news is that projects no longer fail at the 90% rate that we came from in the mid-2000s; but, regardless, despite all we know, all we have studied, all we have researched into why projects fail, project still fail.

Angel and I have discussed this at great length as he and I have equally focused our careers on program and project management. What I admire about Angel and his team, though, is this infectious passion for bucking the status quo; for not accepting that projects have to fail; for how they rise to the challenge of developing professional project management acumen that leads to project success. In the last 10 years, Angel has devoted much of his free time to building ProjectManagers.Org – a free website here PgM's and PM's from all around the world can go to share information, learn about education opportunities and, of course, network.

But, of course, why stop there? So, about 8 years ago, Angel created Courses10.com where people can go to learn about a wide variety of topics – often for free.

About 5 years ago, Angel saw the enormous opportunity AIOps was set to provide and quickly jumped in to learn as much as he could about the topic. Of course, in his selfless way, he couldn't hoard the knowledge he has gained. He needed to find a way to share it. On Courses10.com, you will find an AIOps course which I encourage you to peruse and consume.

Of course, a key question, along this pursuit, is how can we as project managers ensure the success of our organization's AIOps projects. The answer is that there has to be some sort of playbook; a guide, for helping PM's who are experts at managing project, but maybe novice level at AIOps technology. Angel and I are very similar in that we like lists. We like punch lists that you can easily review to know if your project is starting from the right place, iterating correctly on the right track and headed to successful completion.

If we look at the trajectory of AIOps, we rapidly see some key areas where AIOps is solving serious problems that plague a variety of industries. So, how do we encourage monolithic industries to venture into a project that deploys a maturing technology? The answer is checklists! These pages may not answer every single question or layout every single decision a AIOps project team is going to have to make; but, it pulls together the collective knowledge of two seasoned industry experts to provide the cookbook that can help get your initiative off the ground.

This is a great step and I applaud Mssrs. Berniz and Phee-Lip Sim for canonizing their knowledge in this format. I hope it helps you on your projects and I wish you a happy learning!

Best,

Jonathan Goldstein, MBA, CSM, PMP

McKinney, TX

About the Authors

Angel Berniz is an internationally acclaimed digital strategy and growth advisor helping executives transform their Business & Career process through unique digital strategies and transformation roadmaps.

Angel has led IT Strategy & Transformation for thousands of companies, spanning a wide range of areas of expertise. He has amassed a vast variety of proficiencies in this field and simultaneously incorporated them harmoniously on a deep level, driving customers' success in both Digital Business Strategy and Digital Technology Excellence.

His technology expertise spans a wide range of subjects: blockchain, artificial intelligence, machine learning, deep learning, big data & analytics, cloud computing, microservices architecture, internet of things (IoT), robotics, industry 4.0, etc.

In regards to Digital Business Transformation, he makes an analysis of the business value chain and current KPIs and chalks out transformation plans for targeted business goals, and leads digital transformation roadmaps that create tangible business results.

Phee-Lip is a seasoned Technology Operations Leader with more than 20 years of on-the-ground experience. He has helped many projects transition into successful operations and has continued to optimize the performance of these systems. Anchored on strong ITIL fundamentals, he practiced agile and lean ways of working to reduce operational overheads while accelerating time-to-value.

Using technology and human-centered design, Phee-Lip led key business stakeholders to create values fast. Focusing on User Experience, he drove the adoption of Application Performance Monitoring and AIOps to help organizations attain service stability and reliability with significant performance gains.

Thousands of companies leverage Angel Berniz and Phee-Lip Sim's expertise to define and lead their digital strategy, helping businesses improve revenue while reducing cost and risk.

To learn more about the authors, visit their LinkedIn profiles.

Acronyms

BR – Business Requirements

BRS – Business Requirements Specification

COTS – Commercial of the Shelf

ER – Entity Relationship

ER-D – Entity Relationship Diagram

FR – Functional Requirement

FRS – Functional Requirements Specification

FRD – Functional Requirements Document

KPI – Key Performance Indicators

MSA – Master Service Agreement

PO – Purchase Order

RACI – Roles, Accountability, Consulted, and Informed

RAD – Rapid Application Development

RCA – Root Cause Analysis

SOW – Statement of Work

SR – Software Requirements

SRS – Software Requirements Specification

SRD – Software Requirements Document

UR – User Requirements

URS – User Requirements Specification

URD – User Requirements Document

QA – Quality Assurance

QC – Quality Control

AIOps Strategy & Growth

What is AIOps?

AIOps is the new trend in IT Operations. It stands for Artificial Intelligence for IT Operations (previously it was referred as "Algorithmic IT Operations Analytics"). The term refers to IT operations platforms that use Artificial Intelligence. Many of these platforms are cloud services.

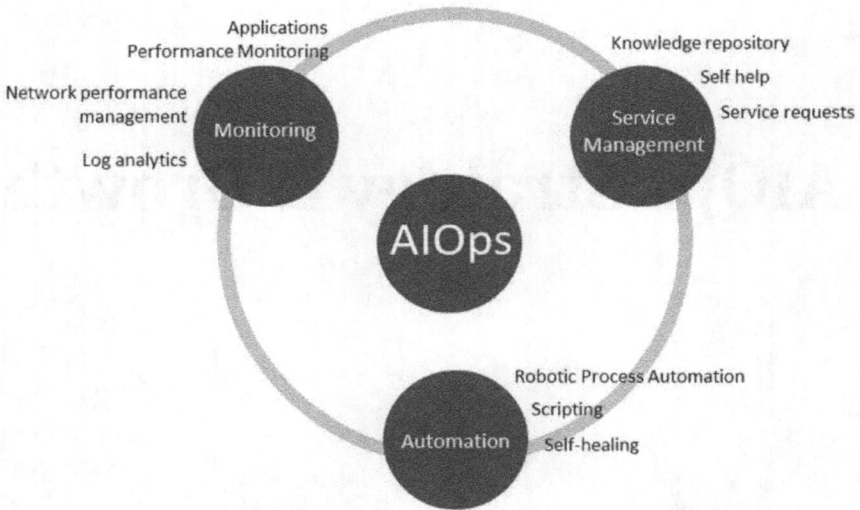

AIOps is the use of advanced algorithms and artificial intelligence techniques for analyzing big data from various IT and business operations tools, to speed service delivery, increase IT efficiency and deliver a superior user experience.

AIOps enables a move away from siloed operations management and provides intelligent insights that drive automation and collaboration for continuous improvement.

In this book, we will focus on the monitoring aspects as the key pillar which enables AIOps.

Monitoring Key Concepts

Availability Monitoring

Availability monitoring is the process of testing the uptime, availability, and average response time of one or more websites to ensure that basic performance and functionality is operating as expected.

It can be done inside your own network or on a global scale leveraging multiple test locations. It is the simplest form of monitoring.

Application Performance Monitoring (APM)

It is different from conventional infrastructure monitoring which primarily captures and reports on hardware performance such as CPU and memory. APM tells you how the application, which sits on top of the infrastructure, is performing by going deep into the code level. It is capable of measuring and evaluating the performance of a software applications' logical and physical stack (from application to infrastructure).

APM stands as the primary means to detect, triage, and remediate on-going performance issues. It provides capabilities such as real-time performance monitoring of applications, ensuring their availability at all times. This is not limited to the applications themselves but also extends to auto-scaling groups and resources.

Apart from availability and performance monitoring, APM supports problems triage leveraging automated and detailed transaction traces, while helping a better matching of application workloads to infrastructure capacity through a real-time analysis of performance bottlenecks over time. Correlating topology changes to application performance to prevent service interruption is yet another important feature of APM.

Recently, it has expanded to include user experience monitoring, which encapsulates capturing of the user journey, reporting of errors and performance of user-triggered activities (click-on-page), as user behavior and experience are becoming more essential. We shall discuss this and more in the Digital Experience Monitoring section.

With huge amounts of data being collected, it is only natural that it has become a big data platform for companies to gain insights into their operations and business. As application complexity increases, APM tools also employ analytics to reduce false-alarms, pinpoint anomalies, and predict problems. It is a boon to organizations to help maintain the highest level of satisfied customer experience.

In addition, most of the APM tools are capable of automatically detecting your entire technology stack end-to-end with the entire application topology visualized in an interactive dashboard. This enables causal dependencies between applications, services, processes, servers, networks, etc., both on premise and in the cloud, to be worked out within minutes.

Digital Experience Monitoring

Digital experience monitoring provides business operations monitoring and analytics for the end to end delivery chain. It ranges from end user applications, its components, and all the way through the infrastructure and backend services. There are two main types of digital experience monitoring: Real User and Synthetic Monitoring.

Real User Monitoring is best witnessed through an integrated set of end-user analytics dashboard, application performance, and infrastructure management services to help enable a bird's eye view of user experience, business transactions, and digital infrastructure.

Synthetic Monitoring is a monitoring technique that executes a series of user activities via a scripted recording of transactions. It is proactive monitoring where it can simulate user activities periodically or frequently to detect anomalies before the actual users perform those tasks in real time.

With digital experience monitoring, developers and IT ops teams can gain deep application and infrastructure visibility, cross-tier correlated insights, and guided problem resolution to ensure the delivery of exceptional digital experiences.

Internet of Things (IoT) Monitoring

IoT is one of the disruptors of technology in digital transformation. As more devices are interconnected, they need to be monitored and synchronized effectively for them to work as designed. With IoT monitoring, you can analyze dynamic systems and keep an eye on the devices to see if they are up and running.

IoT monitoring also bridges the gap between devices and businesses by collecting and analyzing diverse IoT data at web-scale across connected devices and applications.

You can bridge performance gaps by optimizing performance across multiple applications, APIs, networks, and protocols.

You also gain actionable insights to improve customer experience, remediate problems, and maximize IoT opportunities.

Application Programming Interface (API) Monitoring

API monitoring, including Restful API, allows DevOps teams to understand and troubleshoot emerging issues across the API lifecycle and helps to collaborate more effectively. API Monitoring also provides visibility on how APIs are performing alongside with related micro-services.

The growth of APIs/microservices has been exponential since its introduction a mere 10 years ago. This has made API monitoring indispensable, fast-tracking the pace of digital transformation globally.

Metrics gathered at the API service layer are part of the application monitoring system that helps in reaching solutions to a root cause for a given problematic API. Customizable visualizations of API performance metrics help predict service impacts and make sure SLAs are met.

Finally, API monitoring enables oversight into API-driven micro services in direct relation to application infrastructure and backend services.

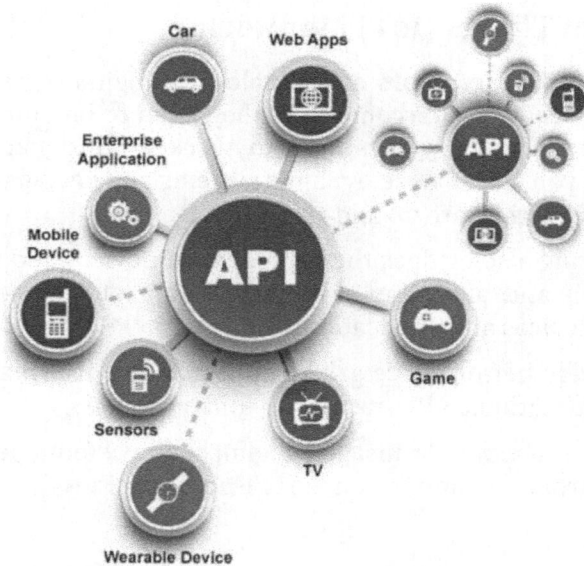

Container Monitoring

Organizations are embarking on to Containerized technology to speed up development efforts and help enable a faster time to market via continuous integration and deployment. *Docker* and *Kubernetes* are the two terms synonymous with Containerization. Docker is a container platform for building, distributing, and running Docker containers while Kubernetes is the most popular container orchestration system for Docker containers. Kubernetes is more extensive than most of the other orchestration systems in the market, hence the popularity. It can coordinate clusters of nodes at scale and in an efficient manner.

However, lack of insights into performance parameters across these complex application architectures can derail the benefits which Containerization can bring.

To mitigate the above problem, organizations are in need of sophisticated monitoring approaches that natively support Containerization and/or similar technologies and don't overburden teams with lengthy configurations to minimize time and effort.

Modern APM better supports Dockers and Kubernetes monitoring with a low-touch, maximum visibility approach, including automatic flow and dependency mapping, adaptive baselining and performance correlation across hosts, containers and applications – in the most complex and demanding distributed microservices architectures.

It has become an essential APM tool for DevOps to rapidly deploy innovative, cloud-based applications by incorporating new approaches in metric capture, correlation, analytics, transaction tracing, and visualization.

Container monitoring can provide complete visibility to tame container chaos and simplify complexity, help optimize performance in constantly changing applications, give insights in massively scalable environments and flexible monitoring that can monitor both containerized and traditional environments together.

Cloud Monitoring

Cloud is the de-facto hosting and infrastructure currently. Hence, it is worth to touch on its key monitoring features. Most of cloud monitoring enables monitoring of at least parts of the components hosted on a cloud environment. It helps identify where the issues are or may arise, thereby reducing mean time to repair. This allows a better user experience by adhering to defined SLAs.

Moreover, this also has features that include quality of service metrics, usage metering, service centric insights, and alerts to help get insights needed to meet SLAs. This enables a boosted up-time while minimizing costs and usage.

Below are the core monitoring tools for the three major cloud providers respectively:

- AWS – CloudWatch
- Azure – Azure Monitor
- Google Cloud Platform – StackDriver

As more Cloud providers are advocating hybrid cloud environments, these monitoring tools help to monitor both public and hybrid cloud-based infrastructure, delivering optimal performance for both. Within each tool, there is a wide variety of sub-tools to help troubleshoot any potential issue that will in turn provide the best usage experience.

When your cloud monitoring solution can manage both on-premises and cloud infrastructure in a unified way, you can seamlessly find and fix problems that might arise during cloud migration.

This cloud diagnostic monitoring feature helps eliminate multiple tools to manage public, private, and hybrid clouds, allowing organizations to reduce costs and provide a seamless monitoring irrespective of the cloud type.

Infrastructure Monitoring

We will briefly go through a few key infrastructure monitoring aspects which are key to AIOps.

Network performance management (NPM) monitors the underlying networking infrastructure to maintain optimal performance levels. It is capable of also troubleshooting and analyzing network performance to prevent bottlenecks. With the introduction of Software Defined Networking, NPM will need to be scaled up to monitor an increasingly diverse and complex set of networking objects and ecosystem efficiently while coping with the exponential growth of bandwidths and data.

Server monitoring helps organizations manage servers across many hardware platforms and a variety of operating systems. It also provides features for quick and easy deployment of servers and can easily manage servers from the smallest environments to more complex levels of enterprise scalability. With intuitive real-time visualization and reporting, one can fix server related problems faster that include server response time issues for any reason. This is only applicable for on premise hosting. Most of cloud platforms monitoring will monitor its computer servers by default.

Database and storage mostly use the vendor proprietary monitoring tools such as Oracle Enterprise Manager for Oracle Database monitoring and Active IQ Unified Manager for NetApp. There are also several options in the market available to monitor NetApp, for example Solarwinds, LogicMonitor, Nagios and OpsView.

IT Operations Analytics (ITOA)

IT Operations Analytics (ITOA) is an approach to collect data related to IT operations and perform analysis on a set of areas of our interest that allows us to understand and make decisions about our IT environment.

Although ITOA tools have emerged over time to bring visibility to IT teams operating in specialized scenarios, they don't operate well in today's complex and heterogeneous environments.

A new breed of ITOA tools better responds to the needs of IT pros responsible for complex applications.

Driven by advancements in machine learning and big-data technologies, ITOA 2.0 deployments include sophisticated data-collection tools, a central data repository and advanced analytics applications to help turn your data into answers for complex business problems. Splunk and Elasticsearch, Logstash & Kibana (ELK) are two of the main ITOA solutions in the market. Splunk is commercially licensed while ELK is the competitive open source alternative.

As many organizations look to ITOA solutions to consolidate monitoring data from different sources, it has good potential to be the steppingstone for AIOps. The downside of this is the consolidated data maybe too fragmented, losing its context, resulting in more efforts to train up big data for AIOps.

Observability

The new buzzword on the monitoring block is observability. Many companies have marketed it as monitoring on steroids and are trying to make quick sales! So, what is observability?

The term comes from the world of engineering and control theory. Observability measures how well the internal states of a system can be inferred from knowledge of its external outputs. Observability is crucial today as modern applications move towards containerized workloads and microservice architectures. The breakup of a system into multiple parts, (complex but distributed systems) which health and performance can then be better measured, make observability possible at this digital age.

According to Will Cappelli, a technology innovation industry veteran (see the References chapter), "with observability's deep connection to causality, monitoring systems must be equipped to provide a causal analysis of the observed system. As we lean into true observability for IT operations practices, practitioners are charged with ensuring the wellbeing of enterprise systems to accurately observe those systems through causality analysis.

While most businesses will reject the idea of "conducting an experiment" on an IT system to prove causality, there is a way to gain insight into causality from the data generated by the system. In fact, system state changes are events that produce a causal relationship when a sequence of those events occur. In turn, with the establishment of causality, we bring about a true understanding of system state changes. This is observability."

This is where AIOps comes into play. Refer to the Auto Root-Cause Analysis and Self-Healing section of AIOps Use Cases for more details.

Service Management & Automation

The other two pillars of AIOps are Service Operations and Automation.

IT Operations Management (ITOM)

ITOM handles the allocation, deployment, monitoring, and managing of IT resources that support digital business processes that are both internal and customer focused. It requires full-stack monitoring and analytics for applications, infrastructure and networks across a variety of cloud types, from mobile to mainframe to help meet the goals of speeding service delivery, while increasing IT efficiency, and delivering a superior user experience.

ITOM is the core orchestrator of AIOps where events for incidents and changes will be analyzed together with the insights gathered from monitoring to manage a particular IT and/or Business service. Rules and patterns can then be configured to orchestrate task automation subsequently.

Automation

The last pillar of AIOps is automation. This is where all the action takes place! Based on the triggers from monitoring and ITOM, actions can be scripted and then executed via run books to perform a particular task. There are many forms of automation, but we will highlight process automation here as it is a software-based approach. The two main categories are:

- Robotic Process Automation (RPA) – automation of a specific process task to deliver tactical efficiencies (e.g. launch a system on screen, check the value of a specific field in the system and then click the submit button if the value is within a particular range). It is typically attempting to automate a specific manual process.

- Business Process Automation (BPA) – automation of an entire business process chain. Sometimes it can be called Digital Process Automation (DPA) in particular if the business process is automated in the context of digital transformation.

It may be relabeled as Intelligence Process Automation (IPA) when artificial intelligence is being applied in the automation execution. Note: these terms are sometimes used interchangeably; hence, clarification to understand the context is necessary.

There is a mixed bag of automation orchestration tools in the market which AIOps systems can leverage. Some of the tool leaders are Chef, Ansible, Puppet, and Terraform. It is worthwhile to evaluate each of their capabilities via proof of value prior to implementation to ensure they're fit-for-purpose.

AIOps Use Cases

IT operations create a heavy volume of data. This typically could range from temperature of a server room/rack enclosure to an inactivity rate of an Application Programming Interface (API). It is also conceivable to obtain information from different layers of the stack. Whenever this data is accumulated, standardized, and broken down, it can turn into a rich source of valuable information.

Here are the primary use cases of AIOps maturing at the point of this writing:

Capacity Planning and Management

While today's cloud capabilities provide dynamic capacity planning, architects still feel that connecting to the correct servers and VM configurations need improvement.

Cloud computing solutions offered by major players such as Amazon (AWS), Microsoft (Azure) and Google (Google Cloud) have a variety of solutions for spinning up VMs by allowing users to choose from a variety of control parameters including but not limited to CPU, Memory, Network, Storage, etc. When business operational tasks in a cloud need scaling, these control parameters are bound to grow over time and increase the complexity.

Processing workloads can be mapped to the exact set of Servers and Virtual Machines by leveraging artificial intelligence capabilities. Looking at the pattern of these workloads, AIOps recommend the right configuration including but not limited to instance types, Storage, IO throughput, and so forth. This will certainly eliminate the uncertainty and challenges that many IT operations face from choosing the right configuration for the right processing workloads thereby reducing the operational and asset management related costs.

Cloud solution offers elasticity to perform business operations where applications can be scaled-out, scaled-in, or even scaled-up. These scaling solutions are either responsive or proactive. In responsive scaling, assets are balanced based on parameters such as CPU utilization or based on the work in a queue. It is also important to plan the scaling tasks to trigger activities per a defined schedule.

With the advent of AIOps, users might depend on prescient scaling where the operation intelligently alters itself based on authentic information. This intelligence helps build a next generation capability such as self-reconfiguration using present and past usage patterns, minimizing human error and optimizing cost.

Predictive scaling takes elasticity, a much needed feature on cloud IAAS, to the next level where scaling happens intelligently and where there will not be any guidelines and configuration settings to empower flexibility. This would determine the ideal assets required through continuous monitoring.

AIOps scope isn't limited to computing assets alone. By leveraging the power of computer-based intelligence, Capacity and System will also be affected.

Effective usage of assets is guaranteed by leveraging computer-based intelligence. Should there be lower IOPS (*Input/output Operations per Second*) or if the capacity is full, this system alerts the users to take appropriate action. Asset piling is diligently balanced by including new volumes proactively through prescient investigation.

Noise Reduction & Anomaly Detection

AIOps can recognize issues in near real time by correlating data from the different layers of the technological stack. Automated application of machine learning noise-reduction algorithms help take away noise produced by huge volumes of data which may be duplicated, redundant, or negligible.

Application and Operational logs that are arranged by time enable propelled machine learning calculations to discover exceptions. AIOps can precisely pick the anomalies by pinpointing the real source that will help IT groups to perform effective underlying driver investigation in approximate real-time.

Anomaly recognition is one of the best use cases of AIOps that can detect potential blackouts and disturbances related to application infrastructure. It then leads to the topic of auto root cause analysis and self-healing.

Auto Root Cause Analysis & Self-Healing

Root Cause Analysis is a complex time-consuming task which involves huge amount of historical data. AIOps is trained to analyze historical actions taken which led to the failure in order to identify the root cause. The more primitive technique uses a rule-based system to identify many common failure scenarios. Machine learning (ML) is now being used extensively.

ML models are trained and used to generate scores characterizing the confidence level of an accurate diagnostic of a root-cause. They learn what successful actions look like over time and use this training to identify failed actions. The system then adapts to changing application behavior while attempting to identify the actual root causes more effectively.

Application performance analytics are the primary starting point as it provides an application-aware context which is crucial to providing the intended business service. Context provides the setting or circumstances associated with application or service events. It then draws out the business impacts and other important use cases across the service lifecycle based on the root cause analysis.

Once the possible root cause has been identified, suggestions for resolving the failure are automatically generated based on historical actions taken. The system then continues to learn from this action-result feedback loop to improve its recommendation capability going forward.

Once the action has been selected (manually or automated), automation service for the fix will kick-in to self-heal the system. This is a classic example on how AIOps leverage the capabilities of monitoring, operations management, and automation services to deliver its potential.

Risk Mitigation via Predictive Maintenance

AIOps uses data-driven analytics to optimize equipment upkeep. This is predictive maintenance. With the explosion of IoT devices, whether on the factory floor or outdoor field sensors, predictive maintenance will drive significant reduction of repair time and unplanned downtime. This will result in extension of equipment lifespan, savings on maintenance cost and, more importantly, in the improvement of health and safety for those who deal with the devices.

AIOps leverages the IoT-generated data stream such as sensor data, error events, temperature and more. Monitoring data in real-time, it uses machine learning to learn normal behaviors. It then picks up anomalies, runs diagnostics and then makes recommendations to upkeep reliability and optimize cost.

AirAsia, the world's top low-cost airline, is one of the best examples of how AIOps can be used not only in predictive maintenance but also in operations management, ground operations, and customer service. The airline collects data from more than 24,000 sensors and uses the data to prevent anything from going wrong. Costs and efficiencies drive everything. "The quicker the aircraft gets back in the air, the more it benefits our customers and us from a cost perspective," the CEO of AirAsia Aireen Omar explained.

Some vendors have made predictive maintenance specific use cases available. For example, Splunk Essential for Predictive Maintenance has been created for maintenance reliability leaders and engineers who are looking for ways to improve and optimize their current preventive maintenance program.

AIOps Case Studies

Case Study I: Auto-Discovery

A global semiconductor company deployed an AIOps platform with Application Performance Management's auto-discovery feature to reduce Mean Time to Repair (MTTR), increase visibility, and gain new insights across its worldwide data centers.

Challenge

Limited visibility of hybrid IT infrastructure and lacking automatic discovery of configuration item (CI) relationships. Key issues:

- There is a lack of dynamic and complete IT infrastructure inventory information.
- Manual tracking and update of relationships between CIs in ServiceNow configuration management database (CMDB) is time-consuming and slow updates are meaningless because relationships become outdated quickly.
- Issues with IP address and CI parameter duplication are difficult to detect and resolve due to lack of visibility.
- There is a lack of automated correlation between application services and infrastructure entities.

Solution

Agentless auto-discovery enablement:

- Discovery of tens of thousands of devices in real time.
- Visibility into data center infrastructure entities and their relationships with real-time update capability.
- Automatic and dynamic updates of CI relationships across entities in ServiceNow CMDB.
- Quick detection of IP and CI parameter duplications in data centers by device type and IP address.
- Discovery of application services and their relevant infrastructure enable efficient migration and IT optimization.

Results

Out-of-the-box CI relationship discovery and CMDB update delivers:

- Reduction of IT costs via resource optimization.
- More powerful ServiceNow IT service management capabilities with automated out-of-the-box CI relationship ingestion.
- More efficient IT auditing and compliance with the use of relationship mapping.
- Better clarity into infrastructure landscapes and their relationships for faster ticket verification and root-cause analysis.
- Dynamic topology maps for all global data centers with real-time updates.

Case Study II: Root Cause Analysis

A leading healthcare company pinpoints the root cause of SAP business process issues in minutes using AIOps to automate root-cause analysis by auto-discovering hybrid infrastructure entities and their corresponding SAP business applications.

Challenge

The Chief Information Officer (CIO) had challenges in delivering required business service level agreements (SLAs) for critical SAP business processes due to the following:

- **Impaired visibility in hybrid IT:** Dynamic and up-to-date inventory of IT infrastructure entities supporting enterprise resource planning (ERP) business processes is lacking.
- **High mean-time-to-repair (MTTR):** Operational data is siloed, existing in disparate domains of the SAP ERP stack. Manual correlation is time-consuming and incomplete, and it quickly gets out-of-date. Automated correlation between business, application, and infrastructure is required.
- **Failed order processing:** Delayed order processing and failed integration between interfaces.

Solution

AIOps' solutions which are application-aware (in this case seamless integration with SAP) coupled with matured capabilities in components correlation, anomaly detection and auto root cause analysis enable:

- **Smart-auto discovery:** Smart auto-discovery of multi-domain application and infrastructure entities keeps inventory up to date. Topology mapping provides granular relationships between physical, virtual, and logical compute, storage, and network entities.
- **Dynamic application discovery and dependency mapping:** Dynamic application discovery and dependency mapping of SAP components and non-SAP applications with underlying infrastructure and an overlay of operational health and performance data collected from various sources.
- **Scalable and out-of-the-box integration with SAP solution manager:** Scalable integration with SAP Solution Manager for business, operations, and infrastructure metrics and time-series event correlation enables quick identification of root cause for ERP users and business key performance indicator (KPI) issues.

Results

Rapid root cause analysis, infrastructure planning, auditing, and compliance:

- Reduced MTTR for SAP ERP KPI issues from hours to minutes.
- Automated auditing and compliance of IT assets for ERP.
- Optimized infrastructure resource planning and workload management for all SAP ERP applications.
- Increased application assurance, improved ERP system uptime, transaction uptime, and user satisfaction.
- Faster value generation from IT to business, and precise business impact analysis ensures business continuity.

Business Analysis

AIOps Business Analysis

As with all technology projects, the success of an AIOps project depends on how well it can fulfill business needs. Hence, accurate definitions of business use cases underpin the success of an AIOps project.

However, as AIOps is still in its infancy many of its use cases are still being discovered. Thus, it is crucial to frame it flexibly to deliver immediate business requirements while positioning it viably in your enterprise architecture in anticipation of its future expansion.

The other two major critical success factors in delivering to business expectations are:

- Organizational capabilities – ensuring that the organization is adequately skilled to run AIOps platforms and initiatives. AI skillsets within the operations domain are difficult to come by. There will not be an IT Hero who knows it all. Rather, a multi-disciplined matrix team must be pulled together to overcome these capability gaps.

- Time-to-value – as AI/machine learning requires significant efforts and time to train, delivering in an agile fashion is vital. Identify low-hanging fruits for immediate results to be achieved while building confidence over time to gather more executive support.

Checklist for AIOps Software Requirement Specifications

An AIOps software requirements specification (SRS) is a document that captures complete description about how the AIOps solution is expected to perform. It is usually signed off at the end of requirements engineering phase.

Qualities of SRS

- Correct
- Unambiguous
- Complete
- Consistent
- Ranked for importance and/or stability
- Verifiable
- Modifiable
- Traceable

Code	Description	Self-check	Peer review	Comments
SRS.1	Are all the AIOps functional capabilities captured?			
SRS.2	Are the analysis models defined in SRS?			
SRS.3	Are the AIOps functional data flows specified including the sources and destinations?			
SRS.4	Is the AIOps SRS free from			

	logical conflicts?			
SRS.5	Is the AIOps SRS document's acronyms and abbreviations consistent?			
SRS.6	Are the AIOps requirements achievable with current available technology?			
SRS.7	Are all AIOps requirements verifiable by some means?			
SRS.8	Are the AIOPs requirements testable?			
SRS.9	Are symbols and notations used in AIOps SRS consistent?			
SRS.10	Are UI controls (Buttons, forms, reports, screens, etc.) included in AIOps SRS?			
SRS.11	Is the AIOps reference to traceability matrix included?			
SRS.12	Is the confidentiality status (Project/Circle/Company) included in the header/ footer of the AIOps SRS?			
SRS.13	Does the AIOps SRS provide enough information about the kind of user interfaces that the different users may use?			

SRS.14	Does the document have good clarity about the AIOps goals, objectives of the project, and the benefits that potential users' software will derive from it?			
SRS.15	Are the AIOps requirements in the SRS traceable to requirements in the URS?			
SRS.16	Does the AIOps SRS provide clear information about potential users of the system?			
SRS.17	Does the AIOps SRS provide enough details about each functional requirement (e.g. operating specifications, external interface specifications, performance, key attributes, other requirements, etc.)?			
RS.18	Does the AIOps SRS provide enough details about the external hardware, software and communication interfaces?			
SRS.19	Does the AIOps SRS provide enough details about the human interfaces?			
SRS.20	Does the AIOps SRS provide enough details about the desirable attributes			

	(reliability, maintainability, portability, safety, security, privacy, etc.) that have been flagged in the URS as relevant and important?			
SRS.21	Does the AIOps SRS have enough details about the reports to be generated by the system (Layout, content, periodicity, distribution, presentation details, etc.)?			
SRS.22	Does the AIOps SRS provide enough details about the performance requirements and is the environment and load for each performance requirement clearly specified?			
SRS.23	Is system performance achievable within the imposed constraints?			
SRS.24	Does the AIOps SRS provide enough details about all key assumptions, dependencies, and risks?			
SRS.25	Are there any ambiguities and design constraints?			
SRS.26	Can the AIOps SRS be used to directly trigger design activities (i.e. does it contain all the information that may			

	be needed during design)?			
SRS.27	Does the AIOps SRS include any design details?			
SRS.28	Does the AIOps SRS define any system test details?			
SRS.29	Does the AIOps SRS define system functions and requirements and are they easily derived from the enclosed analysis outputs?			
SRS.30	Are all AIOps requirements clear, complete, verifiable, and consistent with other requirements?			

Checklist for AIOps User Requirements Specifications

Code	Description	Self-review	Peer review	Comments
URS.1	Are the AIOps user requirements prioritized?			
URS.2	Are AIOps user requirements defined in single sentences?			
URS.3	Does the AIOps URS provide complete details about the objectives of the project and the benefits that potential users will achieve from it?			
URS.4	Is the role of AIOps Service provider clear in terms of scope?			
URS.5	Have all the AIOps external interfaces been defined?			
URS.6	Does the AIOps URS provide complete details on what kind of users will be using the system?			
URS.7	Is there enough information about the kind of AIOps user interface that is required?			
URS.8	Does the AIOps URS define the			

	customer constraints?			
URS.9	Does the AIOps URS define all the assumptions, dependencies and risks and have they been identified?			
URS.10	Have the AIOps unstated requirements been identified?			
URS.11	Does the AIOps URS define the statutory and regulatory requirements related to the project if any?			
URS.12	Are the provided AIOps requirement IDs clear and consistent?			
URS.13	Are the AIOps requirement descriptions documented completely?			
URS.14	Is the confidentiality status updated in the header/footer of the AIOps URS?			
URS.15	Has the technological feasibility of the AIOps project been established?			
URS.16	Is each requirement in the AIOps URS traceable to the customer?			
URS.17	Can the AIOps URS be directly used to prepare the SRS?			
URS.18	Are the AIOps requirements the same as what is stated earlier? If			

	not, has the effect on project parameters (effort, schedules, costing etc.) been estimated and documented?			
URS.19	Is there enough information in the AIOps URS about quality attributes (e.g. reliability, maintainability, portability, safety, security, privacy, etc.) that are likely to be considered as relevant and important by customers?			

Checklist for AIOps Use Case Modelling

Code	Checkpoints	Self-Review (Yes/No/NA)	Peer Review (Yes/No/NA)	Comments
UCM.1	Is the system functionality understandable by reviewing the AIOps use case model?			
UCM.2	Are there any actors that are not defined in the AIOps use case model?			
UCM.3	Does every AIOps use case have at least one actor?			
UCM.4	Does the AIOps use case name describe the behaviour of the use case?			
UCM.5	Are the AIOps use cases very			

	complex?			
UCM.6	Can a functional requirement be mapped to at least one AIOps use case?			
UCM.7	Are there any indirect actors who influence the AIOps use case model?			
UCM.8	Are all the actors clearly described in the AIOps use case?			
UCM.9	Are all the actors connected to the right AIOps use cases?			
UCM.10	Does the communication sequence between actor and AIOps use case conform to user's expectations?			
UCM.11	Is the division of the model			

	into use case packages appropriate?			
UCM.12	Have all the AIOps use cases been identified?			
UCM.13	Is the AIOps use case complete in terms of its goals?			
UCM.14	Do all the AIOps use cases lead to the fulfilment of exactly one goal for an actor, and is the goal evident from the use case name?			
UCM.15	Are the descriptions of how the actor interacts with the system in the AIOps use cases consistent with the description of the actor?			
UCM.16	Is AIOps use case verifiable			

	and/or testable?			
UCM.17	Is the goal of the AIOps use case mentioned clearly in the description?			
UCM.18	Have all the AIOps use cases been identified?			
UCM.19	Is the AIOps use case complete in terms of its goals?			
UCM.20	Are there any superfluous use cases (i.e. AIOps use cases that are outside the boundary of the solution, do not lead to the fulfilment of a goal for an actor or duplicate functionality described in other use cases)?			
UCM.21	Is the AIOps flow of events described			

	with concrete terms and measurable concepts with a suitable level of detail?			
UCM.22	Are there any variations to the normal flow of events that have not been identified in the AIOps use cases?			
UCM.23	Are the triggers, starting conditions, for each AIOps use case described at the correct level of detail?			
UCM.24	Are the AIOps preconditions & post-conditions described with the			

	correct level of detail?			
UCM.25	Does each event in the normal flow of events relate to the goal of its AIOps use case?			
UCM.26	Has the include-relation been used to factor out common behavior?			
UCM.27	Does the behavior of AIOps use case conflict with the behavior of other use cases?			
UCM.28	Are complex AIOps use cases split up?			
UCM.29	Are expected input and output correctly defined in each AIOps use case?			

UCM.30	Is the output from the AIOps solution defined for every input from the actor, both for normal flow of events and variations?			

Checklist for AIOps Use Case Specification

Code	Checkpoints	Self-Review (Yes/No/NA)	Peer Review (Yes/No/NA)	Comments
UCS.1	Is a common naming convention followed for the AIOps file?			
UCS.2	Should the flow of events of one AIOps use case be inserted into the flow of events of another? If so, has this been modeled as an "extend" relationship to the other use cases?			
UCS.3	Are complex AIOps use cases split up into multiple use cases?			
UCS.4	Is the Role and purpose of the AIOps use case clearly documented?			
UCS.5	Are "Out of scope" items mentioned in the AIOps use case?			

UCS.6	Are the AIOps pre-conditions documented?			
UCS.7	Is the Chronological order of the AIOps preconditions correct and complete?			
UCS.8	Are all possible AIOps outcomes documented?			
UCS.9	Are AIOps outcomes for the alternate and exceptional flows documented?			
UCS.10	Are the AIOps Primary and Secondary Actors identified?			
UCS.11	Are all applicable actors of AIOps requirements / vision documents identified?			
UCS.12	Does the communication sequence between actor and AIOps use case conform to the user expectations?			
UCS.13	Are relevant actors connected to the AIOps use cases?			
UCS.14	Have all possible events that trigger the AIOps use case			

	been described?			
UCS.15	Are AIOps outcomes of the alternate and exceptional flows documented?			
UCS.16	Is each AIOps user action and corresponding system reaction/response clearly documented?			
UCS.17	Is data exchanged between the AIOps actor and the system identified?			
UCS.18	Is documentation available to describe how the AIOps use case begins?			
UCS.19	Are there correct references to AIOps Business Rules, Messages, Relationships, Integration Points, non-functional/other Requirements, Alternate and Exceptional flows?			
UCS.20	Is the order, priority, or limitations of the AIOps use case documented?			
UCS.21	Is documentation available to			

	describe how the AIOps use case ends?			
UCS.22	Is a standard and appropriate name given to the AIOps use case specification?			
UCS.23	Is the AIOps Alternate Flow correctly stated?			
UCS.24	Is AIOps documentation available to describe how the alternate flow ends?			
UCS.25	Are the AIOps Alternate Flows numbered (AF1.1, AF1.2)?			
UCS.26	Have additional detours that stem from the AIOps alternate flows been described as another alternate flow with a reference back to the original alternate flow? Have been described Requirements, Alternate and Exceptional flows?			
UCS.27	Are the AIOps Exceptional Flows correctly stated?			
UCS.28	Is documentation available to			

	describe how the AIOps Exceptional Flows end?			
UCS.29	Are AIOps Exceptional Flows numbered (e.g. EF1.1, EF1.2)?			
UCS.30	Is the Expected input and output of this AIOps use case correctly defined?			
UCS.31	Are all events defined in the use case conforming to the goals of the AIOps use case?			
UCS.32	Are there enough details provided in the AIOps use case to start creating UI, design, and test cases?			
UCS.33	Are all business rules relevant to the AIOps use case been addressed?			
UCS.34	Are AIOps data validation rules clearly described?			
UCS.35	Are all AIOps assumptions documented clearly?			
UCS.36	Are the steps at which the AIOps assumptions are made marked/traced?			

UCS.37	Are all AIOps integration points with other projects and external system been identified and documented?			
UCS.38	Have standard names been given to the AIOps Use Case Specification?			
UCS.39	Are external AIOps systems documented?			
UCS.40	Are all the "included" AIOps use cases referred to here?			
UCS.41	Are all the "extended" AIOps use cases referred to here?			
UCS.42	Are all data items that are referred to in this AIOps use case listed and defined?			
UCS.43	Are all AIOps data items that are manually or automatically populated identified?			
UCS.44	Are "record" locking scenarios identified as part of a given AIOps flow?			
UCS.45	In an alternative flow, is Expected			

	AIOps system response documented if an actor tries to access a locked "record"?			
UCS.46	Are all AIOps project issues/open items assigned to owners with specific target dates?			
UCS.47	Are all possible AIOps Error and Warning messages numbered and documented with their applicable warning/error situation/conditio n?			
UCS.48	Has Traceability to AIOps requirements been completed?			

Architecture

AIOps Architecture

9 out of 10 clients we spoke to anticipated that AIOps architecture will be very complex. That is true, but simplicity plays a key role in organizing a complex AI landscape.

We strongly recommend the following features when you design your AIOps Architecture, and keep it simple:

- Compute and technology resources that are scalable, elastic and resilient

- Contextual analytics and optimized ML algorithms capabilities which are easily extendable

- Platform which is optimized for (massive) data-drive development

- Cloud (modern application)-native support for server-less, event-driven and microservice features

- Secured with built-in governance

- Extensible integration using technology such as RESTful API

- Automation-centric (automation first)

Checklist for AIOps Design Review

Code	Checkpoints	Self-Review (Yes/No/NA)	Peer Review (Yes/No/NA)	Remarks / Details
DR.1	Have all mandatory AIOps diagrams (Use Case, Class, Collaboration, Sequence, Deployment and Component diagrams) been included?			
DR.2	Have all the required AIOps I/O formats (Forms, Reports, Screens, etc.) been enclosed?			
DR.3	Has the reference to the traceability matrix been included?			
DR.4	Have all AIOps assumptions and dependencies been explicitly stated? Are these realistic and reasonable?			
DR.5	Does the AIOps analysis/design depend upon any assumptions made about the system that have not been explicitly stated? Are such assumptions acceptable?			
DR.6	Has the confidentiality status (Project/Circle/ Organization) in the footer of the AIOps design document been			

	updated?			
DR.7	Are the conventions followed consistent with the methodology adopted?			
DR.8	Is complete information about the target AIOps system environment being captured?			
DR.9	Is the development environment fully compatible with the target AIOps environment? If not, is there a viable plan for porting development outputs to the target system?			
DR.10	Have all AIOps design constraints been identified, and are they acceptable?			
DR.11	Are the AIOps design constraints consistent with SRS?			
DR.12	Does the deployment diagram represent a good overview of the proposed AIOps system?			
DR.13	Is there complete information available about AIOps interfaces with different external systems? (It is mandatory to know at least the following: identity of the system, type of interface, contents, media,			

	formats, timing, protocols, etc.)			
DR.14	Have the dependencies/ connections/ communications among the AIOps system - components outlined in the SRS been translated into appropriate interfaces (packages/sub-systems)?			
DR.15	Are the concurrency needs identified and outlined in the AIOps design?			
DR.16	Has an attempt been made to simplify all AIOps interfaces?			
DR.17	Is there a correspondence between SRS requirements and use cases identified in the AIOps use case diagram?			
DR.18	Is the description of each AIOps use case complete, consistent, and correct? Has an effort been made to include existing components (developed for other projects)?			
DR.19	Is each identified use case being represented by one or more classes in the AIOps design document?			

DR.20	Is each class identified in the design document connected to an AIOps use case?			
DR.21	Have the AIOps classes been named as per naming conventions?			
DR.22	Is there complete information available about the AIOps class hierarchies?			
DR.23	Are the collaborations among the AIOps classes identified and represented?			
DR.24	Is there complete information about every AIOps interface with external systems? Can the data structures be linked with user reports and screens?			
DR.25	Have the size and composition of AIOps data structures been estimated? Have provisions been made to guard against overflow?			
DR.26	Has an AIOps data dictionary been established? Can it be used to define data design? Will it facilitate impact analysis for a change?			
DR.27	Does the design clearly specify AIOps data back-up and schedule? Does it specify procedures for			

	restoration?			
DR.28	Does the AIOps deployment diagram represent a good overview of the proposed system?			
DR.29	Is complete information available about AIOps interfaces with different external systems? (It is mandatory to know at least the following: identity of the system, type of interface, contents, media, formats, timing, protocols, etc.)			
DR.30	Have the AIOps dependencies/ connections/ communications between the system - components outlined in the SRS been translated into appropriate interfaces (packages/sub-systems)?			
DR.31	Are the concurrency needs identified and handled in the AIOps design?			
DR.32	Has an attempt been made to simplify all AIOps interfaces?			
DR.33	Is there a connection between AIOps SRS requirements and use cases identified in the use case diagram?			

DR.34	Is the description of each AIOps use case complete, consistent, and correct? Has an effort been made to include existing components (developed for other projects)?			
DR.35	Is each identified AIOps class in the design document mappable with a use case?			
DR.36	Are the naming conventions of AIOps classes consistent?			
DR.37	Is there complete information available about the AIOps class hierarchies?			
DR.38	Are the collaborations among the AIOps classes identified and represented?			
DR.39	Is the number of messages passed among AIOps classes within reasonable limits?			
DR.40	Is there complete information about every AIOps interface with external systems? Can the data structures be linked with user reports and screens?			
DR.41	Are the size and composition of AIOps			

	data structures estimated? Have provisions been made to guard against overflow?			
DR.42	Does the AIOps design specify data backup, restoration policies, and schedule?			
DR.43	Are AIOps storage requirements estimated (for whatever period has been stated in the SRS)? Are peak loads accounted for?			
DR.44	Have AIOps screens and reports given by the customer been retained without change in the design? If not, is there a good explanation for the change?			
DR.45	Is a separate AIOps interface required for the System Administrator/ Manager? If so, has this been properly defined and documented?			
DR.46	Are the AIOps screens as per customer standards/company standards?			
DR.47	Is there standardization in the following? • Appearance / structuring of screens			

	• Usage of function keys • Availability & appearance of icons (in case of a GUI based application) • Tool tip display • Short cut keys • Navigation style • Help access and display			
DR.48	Is there complete information available about the hierarchical organization of AIOps screens?			
DR.49	Does the AIOps design facilitate a good user experience in terms of easy addition/deletion/mo dification of screens, in terms of the following? • Screen layout • Options on the screen • Menus available • Field layout, etc.			
DR.50	Is complete information available about fields on various AIOps screens (including the type, width, source, editing associated with each field)?			

DR.51	Is there a direct association between AIOps screens and use cases identified in the use case diagram?			
DR.52	Is there consistency in the layout of reports?			
DR.53	Is complete information available about AIOps reports? (It is mandatory to know details such as layout, content, periodicity, distribution, presentation details, etc.)			
DR.54	Does the AIOps design facilitate easy addition/deletion/modification of reports?			
DR.55	Is there a direct association between AIOps screens and processes identified in the process model?			
DR.56	Has every logical data structure been converted into appropriate physical structures in the AIOps design document?			
DR.57	Has the path for AIOps storage of each file/table been clearly specified?			
DR.58	Have AIOps databases and directories been named as per naming conventions?			

DR.59	Is there complete information about AIOps table/record layouts in terms of column/field names, referential integrity details, edit/validation conditions, keys/indexes, etc.?			
DR.60	Has an AIOps structured process been followed in identifying Primary and Foreign keys?			
DR.61	Are AIOps secondary keys used to the minimum possible extent?			
DR.62	Does the AIOps design facilitate modification of table/record layouts?			
DR.63	Is there complete information about AIOps database such as tables associated and objects like views, triggers, stored procedures, data definition language procedures, etc.?			
DR.64	Has an AIOps structured process been followed in defining views and triggers?			
DR.65	Has an attempt been made to reduce the number of AIOps views and triggers?			

DR.66	Does the AIOps design facilitate easy modification or re-definition of views and triggers?			
DR.67	Is complete information about administrative details such as AIOps security and access rights, administrative procedures, record locking, etc. available?			
DR.68	Does the AIOps system provide for different types of users in terms of privilege levels?			
DR.69	Is there complete information about AIOps database/file system interface?			
DR.70	Are the AIOps data elements in a file closely related to each other and relevant to the file they are contained in?			
DR.71	Have efforts been taken to minimize excessive duplication of AIOps variables and data-structures?			
DR.72	Are all AIOps data structures initialized properly?			
DR.73	Are program/method specifications in accordance with the corresponding class descriptions in the AIOps design			

	document?			
DR.74	Are program/method specifications complete and unambiguous? Will it be possible to easily convert the AIOps specifications into code?			
DR.75	Does the programming language or the AIOps tool selected support the data structures and programming constructs used in the specifications?			
DR.76	Have the AIOps method specifications been written in a manner independent of any specific language or compiler?			
DR.77	Has AIOps compound or inverse logic been avoided?			
DR.78	Has a provision been made for AIOps error handling?			
DR.79	Have AIOps exceptional conditions been handled?			
DR.80	Does the AIOps design address all the functional requirements stated in the SRS?			
DR.81	Have all the possible AIOps error scenarios been taken care of in the design?			

DR.82	Is the AIOps design capable of achieving the performance requirements?			
DR.83	Have AIOps key assumptions and dependencies been validated?			
DR.84	Have AIOps risks arising out of failure of assumptions and dependencies been identified? Has the impact of risks been adequately analyzed?			
DR.85	Is the AIOps design maintainable?			
DR.86	Has the AIOps design been documented according to standards? Is the documentation adequate for its purpose? If not, is this justified and reasonable?			
DR.87	Will it be possible to define details of the AIOps class Integration & class tests based on the information available in the design?			
DR.88	Does the AIOps data design facilitate Information hiding?			
DR.89	Have all the paths of AIOps screen and report layouts been defined?			
DR.90	Has an effort been made to identify			

	different types of AIOps errors, failures and exceptions? Have adequate remedial actions been proposed?			
DR.91	Can the final AIOps system be tested and supported with existing standard facilities (hardware, software, tools etc.)?			
DR.92	Is the definition of AIOps system controls and security adequate?			

Platforms, Enablers, and Accelerators

As traditional analytics and the data it is meant to analyze grows more dynamic and complex, the role of analytics as we know it is shifting from a tool that drives decision making by delivering data insights to a role that is driving business processes via both recommending the most appropriate actions to take regarding issue resolution as well as triggering actions to resolve issues in an automatic fashion.

With technology advancing at such a breakneck pace in order to make the lives of IT Ops simpler and more efficient, cloud environments are only getting more complex. They will continue to do so, as data streams from an increasing number of sources with the explosion of IoT, relational databases, CRM and Application log data, just to name a few. I'm talking about a lot of data, both structured and unstructured. Monitoring in real time is critical in order for AIOps tools to alert IT immediately so as to minimize system downtime and/or support anomaly detection.

The ability to monitor from end to end in real time is paramount, so that finally humans are positioned to make proactive critical decisions. The core of AIOps platforms is that you have machines taking over huge chunks of repetitive work that sometimes takes data teams hours. This takes the burden off from human hands and puts the humans in control at the end, when the most vital decisions can make or break an organization's IT operation.

Before you go out and make the decision to purchase an AIOps platform, you should ask yourself these 4 questions:

1. Does the Platform Monitor and Detect IT Issues in Real Time?

The area surrounding the term "real time" is a bit cloudy (no pun intended), and true real time refers to the updates or frequency of retrievals of data points in order to present new information where it feels instantaneous. Universal standards put this time at a second. This translates to the time between when a data point is introduced into the monitoring systems and the creation of that data point (alert, event, metric, etc.) This timing should be one second or less.

2. Can the Platform Analyze Historical Data?

While ITOA focuses on historic data, many AIOps platforms provide the ability to ingest the plethora of historic data in addition to real time data. You want a platform that can harness the power that previous customer data provides along with other sources, and can provide you with data-driven insights that will inform the organization on the best path to resolution.

3. How Soon Can I Realize Value from an AIOps Platform?

You'd like to purchase a platform that uses the right mixture of algorithms and methods that leaves no environmental stone unturned in uncovering any anomalies for example. Some of this mixture may require longer term analysis, but many of these anomalies should be detectable within a short time period. Detection quality should improve over time as you feed data that will empower learning and that the system can analyze over longer time periods. Feeding the platform historical data is a big part of giving the system what it needs to learn, so it can make the most appropriate and quickest decisions.

4. Will the Platform barrage my IT staff with Alerts?

You want to implement a platform that will intelligently reduce the number of incidents which is the root cause of alerts. A platform that utilizes an optimal trigger for alerts should use machine learning to prioritize incidents based on crowd sourced feedback and the number of incidents the team receiving the alerts is able to handle. This negates a lot of the "boy who cried wolf" alerts that lead to alert overload.

These are just some of the critical questions that come to mind when selecting an AIOps platform. AIOps platforms enhance IT leaders' capability to make an informed decision by contextualizing large volumes of data. The maturity of AIOps primarily focuses on the area of monitoring where it has enhanced the capabilities of Application Performance Management followed by automation and service management. In the table below, we summarize some of the key AIOps platforms on the market and provide a high-level summary of their core features. As almost every IT product vendors are on their journeys to introduce AI to their solutions, we have chosen to focus on domain-agnostic AI platforms here.

Platform	Profile
Moogsoft	Its "Continuous Service Assurance" framework promises to deliver a superior system user experience by: - Robust ingestion of multi-source events - Applying noise reduction - Correlating to detect issues - Identifying root cause - Promoting collaboration - Learning from experience (knowledge recycle)
Big Panda	It is very similar to Moogsoft in terms of values delivery. Big Panda tries to differentiate itself by claiming to be able to reduce IT noise by at least 95% using its "Open Box Machine Learning" while providing transparency, testability, and control of its algorithms. With its pre-trained machine learning models, extensive integrations and a rapid development model, it hopes to deliver results in just 8 – 12 weeks.
Loom System	Loom System's AIOps engine, Sophie, goes a step further by offering an AIOps Playbook to customize organization implementation. It

	comprises a 6-stage process:
	1. On-boarding of data based on the organizational scope
	2. Personalize its AI engine by adding business context to its lexical analysis engine
	3. Provide a closed-loop machine learning feedback
	4. Custom configure user experience alerts
	5. Taylor made dashboards
	6. Engaging your organization via various communication channels

In January 2020, as we were about to publish this book, ServiceNow announced its acquisition of Loom Systems. This is not surprising, and we predict that this trend will continue. While AIOps' capabilities mature, strong performing AIOps vendors will be absorbed by large technology enterprises to help them compete in the AI domain. It is also worthwhile to mention that many large enterprises, such as Cisco, are investing heavily in AI. Cisco has invested in Moogsoft and acquired Perspica. Perspica was then integrated into AppDynamics, a leading Application Performance Management vendor, to become their AIOps engine.

Lastly, to increase the odds of a successful AIOps platform implementation, we recommend focusing on a specific use case that can deliver strong business values. Next, identify the core rules and conditions to lay out the key decision points for AIOps algorithms to be trained based on historical data. Adopt an incremental approach initially and quickly scale it up to other business areas once promising results start to roll in. It takes focus, time, and money but the rewards are compelling once you have the buy-in of the entire organization. A subset of data can only do so much. But if you have the complete set, you will then be able to derive business values easily. This is the theory of the sum being greater than the parts!

Testing

AIOps Testing

Testing in AIOps requires the niche skill set of understanding AI terms like correlation, regression and prediction, data models.

Here we need to consider the following critical points to achieve high quality with respect to AIOps testing:

- Test Environment Setup
- Test Requirements Review
- Test Cases Review
- Test Execution Review

The below snapshot shows an overview of AIOps Testing:

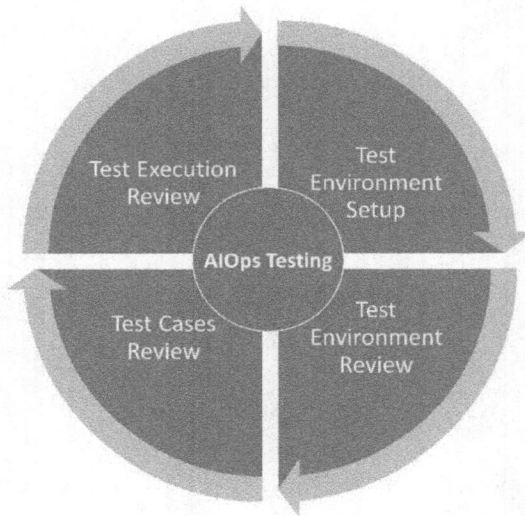

Once this testing base has been established, we would recommend setting up Test Automation which will further:

- Promote the reuse of test cases
- Reduce test cycle time

- Improve the detection of defects

- Expand the test coverage by discovering new testing scope

All of these will reduce the total cost of testing while improving system quality.

The details of test automation are beyond the scope of this book, but at a high level, it will record system testing automatically and re-play it as required. Once the test suite is automated, no human intervention is required. This is a critical success factor for change agility of modern applications where system changes are deployed many times over the week, supporting the automation of continuous integration and continuous deployment (CI/CD) framework.

Many companies are also looking at how AI can improve testing automation. Using AI, the testing tool can learn to identify controls during test execution process. By applying AI learning, images are generalized with image patterns so that they can be used in future testing regardless of the control size, color, and text alignment. The end result is that it enables consistent yet reusable test execution, even on responsive UIs. This is just an example of applying AI in UI test automation.

Checklist for AIOps Test Environment

Code	Checkpoints	Self-Review (Yes/No/NA)	Peer Review (Yes/No/NA)	Remarks / Details
TE.1	Is the AIOps test environment defined by the infrastructure team?			
TE.2	Does the AIOps test environment mimic production environment in terms of hardware/software configuration?			
TE.3	Is the AIOps network type defined *(WAN, Secured LAN, Wireless)*?			
TE.4	Is the AIOps network topology defined *(Star, Hub and Spoke etc.)*?			
TE.5	Are there any AIOps network administration details required?			
TE.6	Is there any AIOps alternative/failover/backup network connection provided to carry on with the testing in case of any emergency/failure?			
TE.7	Are the AIOps Network/System/Server/Client Security level details provided?			

	Are the AIOps firewall details (if any) provided?			
TE.8	Are the AIOps hardware details for Database/Application/ Web Server (or any other servers or architecture components) provided?			
TE.9	Are the AIOps software details for Database/Application/ Web Server (or any other servers or architecture components) provided?			
TE.10	Is the AIOps Configuration (PCs) of client machines specified by the team?			
TE.11	Is the AIOps Configuration (PCs) of client machines defined as per specifications?			
TE.12	Are the access details to the AIOps System/Network/Datab ase/Servers provided *(e.g. ip address for load monitoring, user details for test data verification, etc.)*?			
TE.13	Are the required number of licences for AIOps testing tools provided?			
TE.14	Are there any AIOps Special Security/Access Equipment details required *(Hard Token. Smart Cards, etc.)*?			

TE.15	Are the AIOps configuration management tool licences provided?			
TE.16	Are the AIOps defect management tool licences provided?			
TE.17	Are the AIOps functional test automation tool licences provided?			
TE.18	Are all AIOps in-house automated scripts uploaded?			
TE.19	Is it confirmed that AIOps license expiration date does not end during the test execution period?			
TE.20	Is there any AIOps requirement for remote access of systems? If so, are the access/privileges and details provided?			
TE.21	Are the necessary AIOps backup plans for data/test suites in place?			
TE.22	Does the AIOps test environment comply with security requirements?			
TE.23	Are all required AIOps components deployed in the test environment?			
TE.24	Is the version of each AIOps component mapped with configuration tool checkout list?			
TE.25	Are the differences (constraints) documented?			

	If not, are the differences (constraints) documented?			
TE.26	Are replacement AIOps servers available?			
TE.27	Is the AIOps platform backup procedure available?			
TE.28	Is the AIOps environment refresh procedure available?			
TE.29	If the automated AIOps environment refresh is done, are the scripts available?			
TE.30	Is the AIOps environment usage calendar in place? If not, is the schedule for environment usage communicated?			
TE.31	Are all required automated AIOps tools deployed?			
TE.32	Is user access control specification in place?			
TE.33	Is required AIOps user access provided?			

Checklist for AIOps Test Requirement Review

Code	Checkpoints	Self-Review (Yes/No/NA)	Peer Review (Yes/No/NA)	Remarks / Details
TRR.1	Are the AIOps requirements written in non-technical understandable language?			
TRR.2	Is each characteristic of the final AIOps product described with a unique terminology?			
TRR.3	Is there a glossary in which the specific meanings of each AIOps terms are defined?			
TRR.4	Could the AIOps requirements be understood and implemented by an independent group?			
TRR.5	Are the AIOps requirements written in non-technical understandable language?			
TRR.6	Is each characteristic of the final AIOps product described with unique terminology?			
TRR.7	Is there any index table for AIOps contents?			
TRR.8	Are all the AIOps figures, tables, and diagrams labelled?			

TRR.9	Are all AIOps figures, tables, and diagrams cross referenced?			
TRR.10	Are the possible changes to the AIOps requirements specified?			
TRR.11	Are 'In Scope' and 'Out of scope' AIOps requirements for Testing 'clearly documented?			
TRR.12	Are the types of AIOPs testing as required by the customer clearly documented (e.g. Functional, Performance, Multi-Platform testing, etc.)?			
TRR.13	Are severity/priority of AIOps testing types documented?			
TRR.14	Is the plan for AIOps Integration clear and documented?			
TRR.15	Are AIOps Test entry criteria, Test acceptance criteria, Test Suspension/Resumption criteria known and documented?			
TRR.16	Are AIOps Tools to be used identified and documented?			
TRR.17	AIOps Templates and standards are available for use			
TRR.18	Are there any requirements describing the same object that conflicts with other			

	requirements with respect to AIOps terminology?			
TRR.19	Are there any AIOps requirements describing the same object that conflict with respect to attributes?			
TRR.20	Are there any AIOps requirements that describe two or more actions that conflict logically?			
TRR.21	Are there any AIOps requirements that describe two or more actions that conflict temporally?			
TRR.22	Are all the AIOps requirements traceable back to a specified user requirement?			
TRR.23	Are all the AIOps requirements traceable back to a specific source document or person?			
TRR.24	Are all the AIOps requirements traceable back to a specific design document?			
TRR.25	Are all the AIOps requirements traceable forward to a specific software module?			
TRR.26	Are any AIOps requirements included that are impossible to implement?			
TRR.27	For each AIOps requirement, is there a process that can be executed by either a human or a machine to verify the requirement?			

TRR.28	Are there any AIOps requirements that will be expressed in verifiable terms at a later time?			
TRR.29	Is the AIOps requirements document clearly and logically organized?			
TRR.30	Does the AIOps organization adhere to a defined standard?			
TRR.31	Is there any specific AIOps requirements related to system dependency on other applications and/or platforms?			
TRR.32	Are any specific AIOps requirements identified on interop testing?			
TRR.33	Is there any AIOps requirement identified on system down time during upgrade or fall back operation?			
TRR.34	Are any AIOps System restoration requirements identified/defined in case of current system crash and its restoration time?			
TRR.35	Are the AIOps requirements related to system performance defined?			
TRR.36	Are AIOps requirements related to performance, based upon busy hour processing, defined?			

TRR.37	Are there any Scalability related AIOps requirements identified/defined?			
TRR.38	Are there any external interface related AIOps requirements identified?			
TRR.39	Are there any internal interface related AIOps requirements identified?			
TRR.40	Are there complete security related AIOps requirements documented?			
TRR.41	Are there any encryption related AIOps requirements defined?			

Checklist for AIOps Test Case Review

Code	Description	Self-Review (Y/N/NA)	Peer Review (Y/N/NA)	Remarks
TCR.1	AIOps Test Cases should follow agreed upon standard format.			
TCR.2	The AIOps Test Cases should be complete with respect to Use Cases on which they are based.			
TCR.3	The AIOps Use Case section IDs should be traced back properly.			
TCR.4	Feedback from all previous AIOps reviews should be incorporated.			
TCR.5	AIOps Test data requirements should be mentioned explicitly for each test condition.			
TCR.6	'Expected result' section of each AIOps test case is complete and unambiguous.			
TCR.7	AIOps test steps exist for covering error conditions that can occur. These steps are classified as "negative."			

Code	Description	Self-Review (Y/N/NA)	Peer Review (Y/N/NA)	Remarks
TCR.8	All the messages/error codes specified are correct and same as in the AIOps Use Cases.			
TCR.9	Pre-conditions for executing a test case or a set of AIOps test cases are specified.			
TCR.10	AIOps test cases that are to be executed together (or in a specific order) are specified.			
TCR.11	AIOps Database objects' (e.g. tables, columns, stored procedures, indexes) names, wherever used, are correct and qualified if required.			
TCR.12	AIOps Privileges/Permissions required for executing a test case and validating the results are specified.			
TCR.13	Generic AIOps test cases (e.g. those related to look and feel, validations, navigation, menu & toolbar handling, error reporting, etc.) are identified and documented separately.			
TCR.14	All AIOps queries (questions / clarifications) have been resolved.			
TCR.15	AIOps test cases exist for checking automatic recovery procedures (e.g. checkpoints,			

Code	Description	Self-Review (Y/N/NA)	Peer Review (Y/N/NA)	Remarks
	data recovery, restart, re-initialization).			
TCR.16	Have the AIOps screen layouts been validated?			
TCR.17	Have the checking AIOps cursor navigation and tab order been validated?			
TCR.18	Is there AIOps field validation?			
TCR.19	Has the behavior of each AIOps control (e.g. push buttons, radio buttons, list boxes, etc.) on the screen been validated?			
TCR.20	Has the AIOps processing been validated?			
TCR.21	Check the AIOps report layout (header, Detail, Summary, margins) against the program specification.			
TCR.22	AIOps test cases exist for checking the correctness of data being reported.			
TCR.23	AIOps test cases exist for checking Performance of various operations (e.g. data retrieval and updating, report printing).			
TCR.24	AIOps test cases exist for			

Code	Description	Self-Review (Y/N/NA)	Peer Review (Y/N/NA)	Remarks
	stress testing (high frequency of transaction entry or many simultaneous transaction entries, etc.), if stress testing is planned in project plans.			
TCR.25	AIOps test cases exist for performance (response time, etc.) at high transaction volumes, if performance is an issue.			

Integration and Deployment

AIOps Integration and Deployment

Continuous Integration and Continuous Deployment (CI/CD) should be adopted whenever possible. Beyond being a process, it is also a way of thinking. To make it always deployment-ready, it is suggested to:

- Set up Test Automation for the build (as described in the Testing section)
- Use the same original build through the entire CI/CD pipeline
- Fix problems in the source code (use of a source code version control repository is non-negotiable)

It is recommended to have a one stop-gate between development and production so that you have at least one environment to validate that the release is production-ready. Some projects will have five or more environments along the pipe:

- System testing
- UAT (User Acceptance Testing)
- Staging
- Pre-production
- Production

A well-running continuous delivery pipeline is how easy it is to start it again from the beginning. If a piece of code doesn't work, stop, find out why, fix it, and then start again from the beginning. This will assure consistency which is crucial for CI/CD to work.

To run a successful pipeline, take into consideration the end-to-end process and uncover everything that is done manually. Then, make a plan to automate it. To be truly automated and repeatable, you should also include all changes to the database as part of releases.

Even though your codes are well-tested, a sub-optimal algorithm, failures due to dependencies on other system resources, or other unforeseen conditions can cause a system to behave unpredictably. Within the pipeline, troubleshooting can be tricky. By proactively monitoring telemetry throughout the pipeline, you are more likely to catch problems before they reach production.

Checklist for AIOps Release Planning

Code	Activity	Mandatory (Y/N)	Status (Y/N/NA)	Remarks
RP.1	Has the Hot House Discussion taken place between E2E Release Manager & the following?			
	a. Component SPOCs			
	b. E2E Test Manager			
	c. E2E Delivery Manager			
	d. E2E Vendor Manager			
	e. E2E Configuration Manager			
	f. NDE & Security Team			
RP.2	Has the Hot House Discussion taken place between E2E Vendor Manager & Vendors?			
RP.3	Are the Official Templates being followed in the Program?			
RP.4	Has the E2E Release Manager prepared the E2E Release RAID Register?			
RP.5	Has the E2E Release Manager prepared the E2E Release Plan?			
RP.6	Does the E2E Release Plan contain the following?			
	a. E2E Release Calendar			
	b. Release Components & Their inputs in terms of work packages as per the Release Calendar			
	c. Delivery Strategy for the Release - in case of vendor dependency			
	d. Synchronization Plan for the DEV, SIT, UAT, Pre-PROD and PROD release			
	e. Merging plan for fast track requirements/defect – (Includes Service provider &			

	Vendor Plan)			
	f. Service provider & Vendor Roles and responsibilities of the personnel involved in release work			
	g. Types of environment(s) required, infrastructure details			
	h. Release matrix/ version tree in place			
	i. E2E Release Communication plan – (Includes Service provider & Vendor Plan)			
	j. E2E Release/ Deployment rollout plan – (Includes Service provider & Vendor Plan)			
	k. E2E Release back-up plan – (Includes Service provider & Vendor Plan)			
	l. E2E Release training plan – (Includes Service provider & Vendor Plan)			
	m. Identification of software licensing issues (e.g. obtaining license keys, establishing online validation measures, passwords, dongle, etc.)			
	n. Release Documentation/Templates for the program			
RP.7	Has the E2E Release Plan been reviewed by Component Teams?			
RP.8	Has the E2E Release Plan been reviewed by the E2E Configuration Manager?			
RP.9	Has the E2E Release Plan been approved by the E2E Delivery Manager?			
RP.10	Has the E2E Release Plan been approved by the Quality SPOC?			
RP.11	Has the E2E Release Plan been approved by the Product SME(s)?			

RP.12	Has the E2E Release Plan been approved by the Customer?			
RP.13	Has the approved Baseline Plan been shared with the internal stakeholders and vendors?			

Checklist for AIOps Release Building

Code	Activity	Mandatory (Y/N)	Status (Y/N/NA)	Remarks
RB.1	Has the E2E Solution Architecture been approved by the E2E Lead Solution Architect?			
RB.2	Has the Component team developed code base as per the approved E2E Solution Architecture?			
RB.3	Has the Unit and Component testing been completed by the Component teams?			
RB.4	Has the Integration and Sanity testing been completed by the Component teams?			
RB.5	Has the Component team synchronized their DEV environments with SIT/UAT/PROD on a weekly basis? Has the E2E Delivery Manager ensured the same?			
RB.6	Is the Service provider Configuration Tool updated with Tested Code, Release Notes, Approvals, and other relevant Artefacts?			
RB.7	Has the SPOC audited the Vendor test logs, test results, pass screenshots, and Release documents?			
RB.8	Have the below approvals defined for the Release Notes been obtained? Approvals in sequence as below -			
	a. E2E Solution Lead			
	b. E2E Test Manager			
	c. E2E Configuration Manager			
	d. E2E Release Manager			

	e. E2E Delivery Manager			
	f. Qway SPOC			
RB.9	Has PCA for the release been done by SQA (E2E Config. Mgr. & Quality SPOC)?			

Checklist for AIOps Release Building (Vendors)

Code	Activity	Mandatory (Y/N)	Status (Y/N/NA)	Remarks
RBV.1	Has the Vendor\ Sub Contractor developed a code base as per the approved E2E Solution Architecture?			
RBV.2	Has the Unit and Component testing been completed by the Vendor\Sub Contractor teams?			
RBV.3	Has the Integration and Sanity testing been completed by the Vendor\Sub Contractor teams?			
RBV.4	Has the Vendor\Sub Contractor synchronized their DEV environments with SIT/UAT/PROD on a weekly basis? Has the E2E Vendor Manager ensured the same?			
RBV.5	Is the Service provider Configuration Tool updated with Vendor Tested Code, Release Notes, Approvals and Artefacts?			
RBV.6	Has the Vendor handed over the tested Code base and release notes to Service provider?			
RBV.7	Has the Service provider audited the Vendor test logs, test results, pass screenshots and Release documents?			
RBV.8	Have the below approvals defined for Release Note been obtained? Approvals in sequence as below - a. E2E Solution Lead			

	b.　　E2E Test Manager			
	c.　　E2E Configuration Manager			
	d.　　E2E Release Manager			
	e.　　E2E Vendor Manager			
	f.　　E2E Delivery Manager			
	g.　　Quality SPOC			
RBV.9	Has the vendor sent weekly communication to the E2E Vendor Manager on the release building updates?			
RBV.10	Has PCA for the release been done by SQA (E2E Config. Mgr. & Quality SPOC)?			

Checklist for AIOps Release testing (SIT)

Code	Activity	Mandatory (Y/N)	Status (Y/N/NA)	Remarks
SIT.1	Have the Pre-requisites for SIT defined below been met?			
	a. Is SIT Environment ready, up and running?			
	b. If SIT/UAT test conditions are provided by the Customer, has the Review and approval of SIT/ UAT Test cases by below key stakeholders taken place			
	i. Component Teams			
	ii.E2E Test Manager			
	iii.E2E Vendor Manager			
	iv.E2E Delivery Manager			
	v. Program Manager			
	c. In case of any discrepancies in above, are the same reported and consensus obtained on final SIT/ UAT Test cases from the Customer?			
	d. Has the Test Plan been prepared by the E2E Test Manager?			
	e. Has the Test Plan been approved by the E2E Delivery Manager?			
	f. Has the Test Plan been approved by the Program Manager?			
	g. Has the Test Plan been shared with wider audience?			
SIT.2	Have the E2E Testing Team \ Business Users (if applicable) conducted SIT as per Approved SIT Test cases and UAT Test cases?			
SIT.3	Has the UAT Test cases been			

	used in SIT to reduce the slippage?			
SIT.4	Are the bugs being classified as H/L/M and prioritized by the E2E Testing Team?			
SIT.5	Are daily Test Status Report being sent by the E2E Test Manager to Service provider Stakeholders and Customer?			

Checklist for AIOps Release Testing (UAT)

Code	Activity	Mandatory (Y/N)	Status (Y/N/NA)	Remarks
UAT.1	**Have the Pre-requisites for UAT defined below been met?**			
	a. Is the UAT Environment ready, up and running?			
	b. Have the UAT Test conditions been prepared by E2E Testing Team?			
	c. Have the UAT Test Conditions been approved by the E2E Solution Architect?			
	d. Have the UAT Test Conditions been approved by the Customer?			
	e. If UAT test conditions are provided by the Customer, has the Review and approval of UAT Test cases by below key stakeholders taken place?			
	i. Component Teams			
	ii. E2E Test Manager			
	iii. E2E Vendor Manager			
	iv. E2E Delivery Manager			
	v. Program Manager			
	f. In case of any discrepancies in the above, has the same been reported and consensus obtained on final UAT Test cases from the Customer?			
	g. Has the UAT Test Plan been prepared by the E2E Test Manager?			
	h. Has the UAT Test Plan been approved by the E2E Delivery Manager?			
	i. Has the UAT Test Plan been approved by the Program Manager?			

	j. Has the UAT Test Plan been shared with * wider audience?			
UAT.2	Has the E2E Testing Team\Business Users (if applicable) conducted UAT as per Approved UAT Test cases?			
UAT.3	Are the bugs being classified as H/L/M and prioritized by the E2E Testing Team?			
UAT.4	Are the component teams fixing the bugs as per priority and deploying in UAT Drops 1,2,3, etc. as agreed with the Stakeholders?			
UAT.5	Is the daily Test Status Report being sent by the E2E Test Manager to Service provider Stakeholders and the Customer?			

Checklist for AIOps Release Testing (Operational Readiness Testing)

Code	Activity	Mandatory (Y/N)	Status (Y/N/NA)	Remarks
ORT.1	**Have the Pre-requisites for ORT defined below been met?**			
	a. Is Pre-Prod Environment ready, up and running?			
	b. Have the ORT Test conditions been prepared by the E2E Testing Team?			
	c. Are ORT Test Conditions Approved by the Customer?			
	d. Has the Test Plan been prepared by the E2E Test Manager?			
	e. Has the Test Plan been approved by the E2E Delivery Manager?			
	f. Has the Test Plan been approved by the Program Manager?			
	g. Has the Test Plan been shared with *wider audience?			
ORT.2	Has ORT been conducted by the E2E Testing Team as per Approved ORT Test cases?			
ORT.3	Has the Customer given ORT sign-off & Go-ahead for PROD?			

Checklist for AIOps Release Testing (Performance)

Code	Activity	Mandatory (Y/N)	Status (Y/N/NA)	Remarks
P.1	Has the Team carried out Performance Testing in UAT Env. or different Env. as per customer requirements?			
P.2	Has the customer given Performance Testing sign-off and go-ahead?			

Project Management

AIOps Project Management

We lay down the foundational principles of project management in this section, sharing the key project deliverables. They are intended to be followed sequentially and we strongly recommend the adoption of an agile project management approach.

While you run through these project deliverables, you must bear in mind these following agile way of working which are based on the 12 principles behind the Agile Manifesto:

1. Prioritize what the customer needs through early and continuous delivery of valuable software.
2. Requirements will change, cater for those changes and deliver to create competitive advantage for your business.
3. Break the deliverables down to smaller churns and deliver working software more frequently.
4. Work with the Business daily throughout the project.
5. Motivate your team. Give them the environment and support they need and trust them to get the job done.
6. The most efficient and effective method of conveying information to and within a development team is face-to-face conversation.
7. Working software is the primary measure of progress.
8. Set up cadence which allows the project sponsors, developers, and users to come together at a constant pace to deliver progress sustainably.
9. Continuous attention to technical excellence and good design.
10. Always keep it simple in delivering the project.
11. Set up autonomous, self-organizing teams to produce the best architectures, requirements, and designs.
12. At regular intervals, the team reflects on how to become more effective, then tweaks and adjusts its behaviour accordingly.

Checklist for AIOps Project Initiation

Code	Item	Self-Review (Yes/No/NA)	Peer Review (Yes/No/NA)	Comments
PI.1	Is the preliminary scope defined and project timelines estimated?			
PI.2	Are the project efforts & costs estimated? Have these been reviewed and approved?			
PI.3	Are the risks identified (business, technical, and project management risks)?			
PI.4	Are the acceptance criteria defined? Are the Assumptions in RFP/RFP Response referred/documented?			
PI.5	Are the requirements for Information Security and Business Continuity, if any, are defined and understood?			
PI.6	Are the contractual requirements reviewed and agreed with relevant stakeholders?			
PI.7	Are the critical dependencies identified, discussed, and agreed with stakeholders?			
PI.8	Are the support units and collaboration requirements identified?			
PI.9	Are the application security and infrastructure requirements (also			

	customer site) identified?			
PI.10	Is the project created and approved? Are the project success criteria defined?			
PI.11	Is the project team acquired and visa/work permit/travel requirements identified and initiated?			
PI.12	Is the external procurement identified and initiated where applicable?			
PI.13	Is the handshake from presales to delivery and from sales to deliver complete?			
PI.14	Is project kick off conducted with all relevant stakeholders and MoM (Minutes of Meeting) available?			
PI.15	Are the point of contacts from customer, collaboration and support units identified and their responsibilities recorded?			
PI.16	Is the project criticality assessment conducted and criticality arrived at?			
PI.17	Are reusable components identified required for the project? If yes, have appropriate permissions been taken from all stakeholders for reuse?			

Checklist for AIOps Project Approval

Code	Checklist Item Description	QC Review (Yes/No/NA)	QC Remarks
PA.1	Are PO (Purchase Order)/SOW (Statement of Work)/MSA (Master Service Agreement) available for the project?		
PA.2	Are the project main type & sub-type appropriately identified for the project?		
PA.3	Is the project management category, appropriately identified for the project?		
PA.4	Does the 'opening risk register' exist and has it been filled appropriately and completely with details for mitigation plans, contingency plans, exposure values, etc.?		

Checklist for AIOps Project Handover

Code	Description	Received Documents/	Comments
PH.1	Have you been introduced to the customer?		
PH.2	Have you been given a briefing on client hierarchy, the Organization and the escalations/appreciations in project?		
PH.3	Are all emails pertaining to the release, specific customer communication and important approvals, from customer available in project repository?		
PH.4	Have the details regarding status reporting to customer – Frequency, dashboard, format, etc. pending action items been shared?		
PH.5	Do you know the frequency of conducting customer satisfaction survey?		
PH.6	Have you received the feedback received from the customer along with the status of improvement action plan (if any)?		
PH.7	Do you know the stakeholders of your project?		
PH.8	Are the contract/proposal/concept note along with commitment review checklist available?		

PH.9	Are all the open action points from contract review closed?		
PH.10	Do you know the process of contract linking to your Project?		
PH.11	Is the scope of the project understood clearly?		
PH.12	In case of "First time PM", have you undergone a training on Customer Orientation, Team Management, Communication, Governance, PM's KRAs, Risk Management, BMS etc.?		
PH.13	Do you know the project creation process?		
PH.14	Do you know whether your project uses any tool for project management activities? Do you know how to create a work package in the tool used for project management?		
PH.15	Are you appraised on people, role and their working style and other context w.r.t projects, including the SPOCs in Support functions (finance (for invoicing), HR, Learning, PMO, Quality, RMG, Legal and others)?		
PH.16	Do you know the SLA details and means of tracking it?		
PH.17	Are you fully aware of the review & defect tracking process?		
PH.18	Has the resource plan of the project been shared?		

PH.19	Are the estimations for project/current work package /current release received?		
PH.20	Are you aware of the assets owned by the project?		
PH.21	Do you know the steps to add tasks for entering in the timesheet?		
PH.22	Have you received the project plan document along with schedules?		
PH.23	Do you know the quality goals of your Project? Are they project specific or taken from Org-level goals?		
PH.24	Has the Software Configuration Plan been received?		
PH.25	Has the operational process document applicable for your type of project (If not included in project plan) been received?		
PH.26	Has the risk register been received?		
PH.27	Have all impact analysis details been received?		
PH.28	Do you know how to fill up the metrics reports?		
PH.29	Have all previous metrics data and actions planned or in progress from the metrics analysis been received?		
PH.30	Has Change Register & Change log been received?		

PH.31	Has configuration register's (if no tool is used for configuration management) been received?		
PH.32	Have you received the baseline reports?		
PH.33	Have you received the archival information of previous releases?		
PH.34	Do you know the frequency of various project reports internal & to customer?		
PH.35	Has the latest physical configuration audit/final inspection/SQA report been shared with you?		
PH.36	Have all the managed and controlled documents in the project been shared with you?		
PH.37	Have you received the quality records maintained in the project?		
PH.38	Have you received the last 'Interim Project Closure Report' of the project?		
PH.39	Do you know the latest Internal Audit findings and status of observations being tracked to closure (if any)?		
PH.40	Do you know the status of latest non-conformances raised in the Internal Quality Audits (if any)?		
PH.41	Have you been introduced to security coordinator of your organization?		

PH.42	Are you aware of security requirements for the project?		
PH.43	Are you aware of BIA (Business Impact Analysis), BCP (Business Continuity Plan), and Asset Inventory?		
PH.44	Have any plan of actions for the gaps, process improvement identified discussed, and relevant artefacts been received?		
PH.45	Do you know the current CMMI level of the project and any plan of action?		
PH.46	Do you know where to get standard Service provider document templates?		
PH.47	Do you know where to get best practices, lessons learned, organizational risk database, etc.?		

Transition & Handover

AIOps Transition & Handover

As many organizations are embarking on the DevOps journey, the need for a separate, dedicated Transition and Handover process will become a thing of the past. However, acknowledging that there is still some way to go for organizations to be truly DevOps, we outline the key deliverables for Transition & Handover in this section.

Checklist for AIOps Transition

Code	Item Description	Completed (Y/N/NA)	Remarks/Comments
T.1	Is there a formal presentation of AIOps project processes to the client?		
T.2	Is there a formal presentation of AIOps project application knowledge acquired by the client?		
T.3	Is there a formal presentation of AIOps project knowledge acquired on support process?		
T.4	Is there a formal presentation on how to compile/build/deploy to production and participation in at least one such build-deploy AIOps project process?		
T.5	Has the AIOps service provider attended face-to-face meetings with User/Client group?		
T.6	Are the AIOps SLAs agreed to and signed-off on?		

T.7	Has the AIOps service provider prepared the necessary support documentation as per plan?		
T.8	Has the AIOps service provider team gained sufficient knowledge to take over the shadow phase?		
T.9	Has the AIOps service provider secured successful rating in knowledge tests conducted?		
T.10	Have all AIOps project activities, as agreed and signed-off for knowledge transfer been successfully completed by the Service provider team and have gotten acceptance/sign-off for the same?		
T.11	Is the AIOps escalation process documented, reviewed, and issued?		
T.12	Is the AIOps infrastructure ready in all respects (like hardware, software, network, security, resources, etc.) by the Service provider?		

Checklist for AIOps Transition Handover

Code	Checkpoints	Self-Review (Yes/No/NA)	Peer Review (Yes/No/NA)	Remarks / Details
TH.1	Is there an AIOps project purchase order in place?			
TH.2	Is there a signed AIOps project contract?			
TH.3	Are there estimations for AIOps project/current work package /current release available with the customer?			
TH.4	Has the AIOps project Transition Plan/Charter been created?			
TH.5	Are all the AIOps project review comments are closed and signed-off on?			
TH.6	Has the AIOps project scope, deliverables, Schedules been Signed-off on with the Customer?			
TH.7	Is the AIOps project software configuration plan			

	available?			
TH.8	Has the AIOps Project Manager conducted physical configuration audit as per the SCM Plan?			
TH.9	Is the AIOps project VV&T (Verification, Validation & Testing) Plan available/prepared?			
TH.10	Is all the AIOps project Re-planning information present?			
TH.11	Are the AIOps project risk mitigation plan & tracking sheets in place?			
TH.12	Are the AIOps project issue and query register template for project / work packages/ releases / work under transition ready & available?			
TH.13	Is the AIOps project change register defined & maintained in place?			
TH.14	Is expected AIOps project metric analysis in place and relevant data available?			

TH.15	Is AIOps project configuration register/s (if no tool is used for CM) baseline reports available/prepared?			
TH.16	Is the AIOps project "User Requirement Document" (e.g. User Guide, Cookbook) prepared?			
TH.17	Has the AIOps project URS been reviewed/inspected?			
TH.18	Have necessary AIOps project corrective action(s) (if any) been taken after review of URS?			
TH.19	Are the required AIOps project software licenses available?			
TH.20	Are the AIOps project access cards, privileged passwords for servers/hardware – project servers, configuration management tool, bug tracking tool available?			
TH.21	Are AIOps project user accounts of all resources released from the project disabled?			

TH.22	Are the quality records of the AIOps project available?			

Presales & Bidding

AIOps Presales & Bidding

It is all about value – the value which the Client (Business) is looking for. Besides outlining the artifacts which are important in the bidding process, we will touch on how presales can raise the values which the Business is looking for and nail them with a successful deal.

Business Case

The first step is to develop a compelling business case for implementing an AIOps solution. This includes:

- Identifying the current business challenges and improving inefficiencies
- Opportunities for growth, exploring new customer segment to create entirely new revenue streams.

It is important to translate these cases into clear KPIs which can then be used to measure the success of a proof of value exercise.

Proof of Value (PoV)

You would have certainly noticed that many vendors have replaced Proof of Concept (PoC) with PoV. It is for a good reason – concept does not make a business successful, value delivery does. If success in meeting those Business KPIs can be demonstrated during the PoV engagement, most of the battles will have already been won. Hence, determining that the product/service is fit for the business use cases (KPIs) upfront is the critical success factor here.

Knowledge Transition Post Sales/Execution

Most of the implementation focuses on system deliverables without much consideration on how the solution can be successfully operated post-project execution. This is the biggest mistake as most of the time successful execution of a project does not deliver any immediate business results. It is only after the project team has long gone that the project cost is recovered, and business values are delivered. For a complex AIOps system, you must ensure that the client run team can draw the values the project is intended to deliver. This would need an in-depth knowledge transition and a well laid-out learning path for continuous capacity building. Thus, it is paramount to make sufficient provision for post-implementation support.

Checklist for AIOps Bid Review

	Item Description	Completed (Y/N/NA)	Remarks/Comments
BR.1	Is the AIOps project client an existing customer or a new one?		
BR.2	Is the AIOps project client part of the identified list of target customers?		
BR.3	Will wining this bid help us establish/maintain a long-term relationship with the AIOps project customer?		
BR.4	Will winning this bid further strengthen our overall AIOps project sales strategy and/or help us enter a new domain/market?		
BR.5	Is this new AIOps project domain/market a strategic target for us?		
BR.6	Will serving this AIOps project customer offer us a leadership advantage (e.g. top market		

	player)?		
BR.7	Has AIOps business upside potential been factored in pricing for the deal?		
BR.8	Have AIOps key competitive threats been tackled?		
BR.9	Have relationships been built with key AIOps project client?		
BR.10	Is the AIOps project viable?		
BR.11	Is the AIOps project profitable?		
BR.12	Are there any AIOps project exceptions in pricing?		
BR.13	Are there any AIOps project exceptions in legal?		
BR.14	Are all AIOps project exceptions reviewed and analysed in detail?		
BR.15	Can we comply with all the mandatory legal requirements for this AIOps project?		
BR.16	Can we take AIOps project exceptions to any of the mandatory legal requirements?		
BR.17	Are there alternative measures that can be taken to mitigate the risk of these AIOps project exceptions?		

	And are they viable?		
BR.18	Is the AIOps project client legally clean at the time of bidding?		
BR.19	Is there any other party involved in this AIOps project deal?		
BR.20	Are the AIOps project confidentiality & NDA in place?		
BR.21	Are the AIOps project terms & conditions complete and exhaustive?		
BR.22	Will AIOps project agreement be signed prior to main agreement?		
BR.23	Where appropriate, are there back to back AIOps project contracts in place?		
BR.24	Are the AIOps project teaming agreements in place?		
BR.25	Are the AIOps project alliance agreements in place?		
BR.26	Is AIOps project contract management agreement in place?		
BR.27	Is AIOps project invoicing addressed & covered adequately in the proposal?		

BR.28	Is AIOps project periodic rate escalation (based on COL indices) addressed & covered adequately in the proposal?		
BR.29	Are payment terms addressed & covered adequately in the proposal?		
BR.30	Are late payment charges/interest for delayed payment addressed & covered adequately in the AIOps project proposal?		
BR.31	Are normal working hours addressed & covered adequately in the AIOps project proposal?		
BR.32	Is short term per diem expenses addressed & covered adequately in the AIOps project proposal?		
BR.33	Are travel expenses addressed & covered adequately in the AIOps project proposal?		
BR.34	Are taxes addressed & covered adequately in the AIOps project proposal?		

BR.35	Are delay penalties addressed & covered adequately in the AIOps project proposal?		
BR.36	Are liquidated damages addressed & covered adequately in the AIOps project proposal?		
BR.37	Is performance guarantee/bond addressed & covered adequately in the AIOps project proposal?		
BR.38	Is insurance addressed & covered adequately in the AIOps project proposal?		
BR.39	Are service credits addressed & covered adequately in the AIOps project proposal?		
BR.40	Is gain sharing addressed & covered adequately in the AIOps project proposal?		
BR.41	Is resale of equipment procured by service provider to customer - addressed & covered adequately in the AIOps project proposal?		

BR.42	Are currency fluctuations addressed & covered adequately in the AIOps project proposal?		
BR.43	Are rebates/ discounts addressed & covered adequately in the AIOps project proposal?		
BR.44	Is indemnity addressed & covered adequately in the AIOps project proposal?		
BR.45	Is limitation of liability addressed & covered adequately in the AIOps project proposal?		
BR.46	Are contractual reps & warranty addressed & covered adequately in the AIOps project proposal?		
BR.47	Is the most favoured customer clause addressed & covered adequately in the AIOps project proposal?		
BR.48	Are intellectual property rights addressed & covered adequately in the AIOps project proposal?		

BR.49	Is customer owned IP addressed & covered adequately in the AIOps project proposal?		
BR.50	Is licensing of third-party IP to customer addressed & covered adequately in the AIOps project proposal?		
BR.51	Is confidentiality addressed & covered adequately in the AIOps project proposal?		
BR.52	Is force majeure addressed & covered adequately in the AIOps project proposal?		
BR.53	Is privacy and information security addressed & covered adequately in the AIOps project proposal?		
BR.54	Is termination for convenience addressed & covered adequately in the AIOps project proposal?		
BR.55	Is termination for cause addressed & covered adequately in the AIOps project proposal?		
BR.56	Is compliance with customer policies addressed & covered		

	adequately in the AIOps project proposal?		
BR.57	Is compliance with applicable laws addressed & covered adequately in the AIOps project proposal?		
BR.58	Is escrow of third-party software/code addressed & covered adequately in the AIOps project proposal?		
BR.59	Is addition of affiliates/suppliers/ partners addressed & covered adequately in the AIOps project proposal?		
BR.60	Is dispute resolution addressed & covered adequately in the AIOps project proposal?		
BR.61	Are governing laws addressed & covered adequately in the AIOps project proposal?		
BR.62	Is jurisdiction addressed & covered adequately in the AIOps project proposal?		
BR.63	Is due diligence addressed & covered adequately in the AIOps project proposal?		

BR.64	Is nature of services addressed & covered adequately in the proposal?		
BR.65	Is installation/ implementation addressed & covered adequately in the AIOps project proposal?		
BR.66	Are service levels - master agreement/ relationship level addressed & covered adequately in the AIOps project proposal?		
BR.67	Are service levels - sow/ project level addressed & covered adequately in the AIOps project proposal?		
BR.68	Are critical services addressed & covered adequately in the AIOps project proposal?		
BR.69	Is the transition plan addressed & covered adequately in the AIOps project proposal?		
BR.70	Is the technology plan addressed & covered adequately in the AIOps project proposal?		
BR.71	Are shared resources addressed & covered adequately in the		

	AIOps project proposal?		
BR.72	Is working in multi-vendor environment addressed & covered adequately in the AIOps project proposal?		
BR.73	Are customer satisfaction surveys addressed & covered adequately in the AIOps project proposal?		
BR.74	Are change requests addressed & covered adequately in the AIOps project proposal?		
BR.75	Is transfer of third-party vendor contracts to the service provider addressed & covered adequately in the AIOps project proposal?		
BR.76	Is the relationship governance - steering committee/ executives/ account managers/ project managers addressed & covered adequately in the AIOps project proposal?		
BR.77	Are all AIOps project related risks identified?		

BR.78	Are all the AIOps project risks quantified?		
BR.79	Is there an AIOps project mitigation/avoidance plan in place?		
BR.80	Is the AIOps project cost of all the identified risk prepared and included in the final price?		
BR.81	Has P&L been created for the proposed AIOps project?		
BR.82	Are cash flow scenarios created for the proposed AIOps project?		
BR.83	Are AIOps terms of delivery, warranty, payment, replacements on dead on arrival (in case of hardware) clear?		
BR.84	Where applicable, have all the AIOps procurements related to the proposed project in compliance with the organization policy?		
BR.85	Are there any AIOps exceptions to the organization policy?		
BR.86	Are AIOps approvals sought for organization policy		

	exceptions?		
BR.87	Is the proposed AIOps project requirements complying with STPI policy?		
BR.88	Is the proposed AIOps project requirements complying with SEZ policy?		
BR.89	Will any requirements of the proposed AIOps project have customs clearance issues?		
BR.90	Can the AIOps customs clearance issues be handled?		
BR.91	Is AIOps negotiation with vendors for best price done?		
BR.92	Will there be any issues in the shipping timelines and the proposed AIOps project timelines?		
BR.93	Can we meet the AIOps infrastructure requirements?		
BR.94	Are there any AIOps risks associated to the infrastructure requirements?		
BR.95	Can these AIOps risks be mitigated/handled?		
BR.96	Are the AIOps timelines provided enough to		

	accomplish the staffing requirements?		
BR.97	Can the AIOps staffing requirements for the proposed project be met internally?		
BR.98	Are there any AIOps Immigration/VISA related issues?		
BR.99	Is the AIOps country/location blacklisted?		
BR.100	Do we have presence in the country/location where the AIOps project will be executed?		
BR.101	Are there any AIOps local recruitments required?		
BR.102	Are there any government policies that impact the AIOps local recruitments?		
BR.103	Are there any clearances required for recruitment? If any, Will the timelines for clearance impact the AIOps project execution timelines?		
BR.104	Do we have a local AIOps partner in the country/location for recruitment?		

Technical Writing

AIOps Technical Documentation

The typical considerations for technical writing are provided here. It is prudent to consider other modes of delivery such as video and online learning for making the best impact on education. In this fast-changing pace, documents are obsolete as soon as they are created; hence, they must be updated regularly.

Checklist for AIOps Technical Documentation

	Description	Self-check	Peer review	Comments
TW.1	Do all the AIOps technical tables have the correct heading?			
TW.2	Are all the AIOps technical images displayed correctly?			
TW.3	Are all the cross AIOps technical references linked to the correct text?			
TW.4	Is the AIOps technical content clear and easily understandable?			
TW.5	Are all AIOps technical wordings, data, graphics, or anything borrowed from the literature or from other works is properly referenced and permissions obtained as needed?			
TW.6	Does the AIOps technical document contain sufficient details so that the results can be reproduced by someone else?			
TW.7	Does the AIOps technical			

	document have an introduction stating the purpose of the document and why the document is important?			
TW.8	Is the AIOps technical document spell checked and grammar checked?			
TW.9	Is the AIOps technical document proofread?			
TW.10	Are all AIOps technical tables/figures meaningful and contribute to the document?			
TW.11	Is there AIOps technical information which explains all the parts of the image or figure?			
TW.12	Does the AIOps technical figure reside in the margins?			
TW.13	Is the AIOps programming code defined in one consistent font?			

The Future of AIOps

AIOps Use Cases

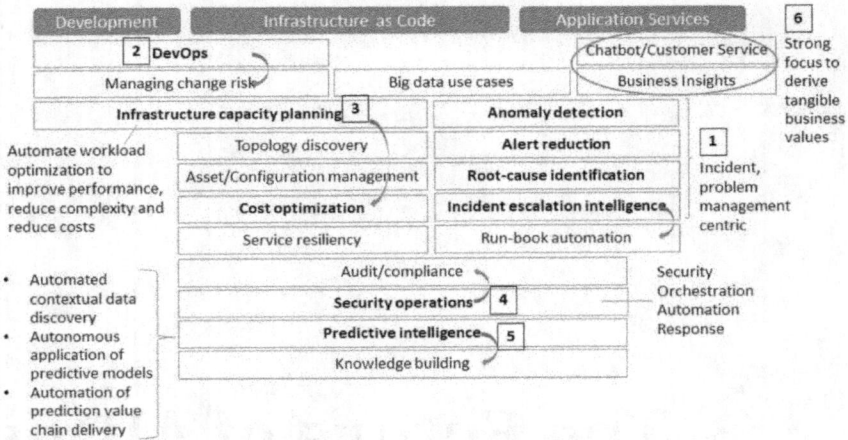

We have analyzed some of the developing AIOps trends and outlined six key themes below.

1. Incident and Problem Management

The majority of AIOps use cases are directed at the incident and problem management of IT service. They typically start with anomaly detection and alert reduction to filter out the noise in the operating landscape. They subsequently improve on accuracy to be able to pinpoint root-causes by providing the context of the technology in play. There is a handful of use cases where AI was used to expedite incident escalation learning from previous incidents management. Ultimately, this leads to run-book automation where self-healing or/and service requests fulfillment are promoted.

2. Continuous Integration and Continuous Deployment (CI/CD)

The focus on CI/CD is on auto-provisioning and management of development stacks. The AIOps solutions tend to employ an infrastructure-as-code approach to CI/CD pipeline automation services. Another use case closely related to CI/CD is the usage of AIOps in managing change and release risk.

3. Infrastructure Provisioning and Management

The other key AIOps trend spotted is on cloud infrastructure management. The primary motivation is to automate platform workload optimization, including scaling up or down on infrastructural resources, to improve performance and reduce costs. The technology has also helped to reduce infrastructure complexity by discovering topology and automate asset/configuration management intelligently. One other use case for AIOps on infrastructure is to provide service resiliency by enabling a just-in-time disaster recovery failover.

4. Security Operations

There is a growing trend for security operations to leverage AIOps for Security Incident Event Management (SIEM) and Security Orchestration Automation Response (SOAR). We predict that the scope will be extending into areas of audit and compliance very soon. As the technology landscape becomes more complex by the day, it is vital to leverage AI to detect and remediate security-related incidents as early as possible to mitigate risk and prevent escalating loss.

5. Predictive Intelligence

There have been predictions of many things, but it has proven to be elusive when it comes to accuracy. However, with the advance of quantum computing, massive computing power coupled with contextual knowledge apply on available datasets, the potential of predictive analytics maybe revived. Below are the few aspects of predictive analytic which we expect to grow as AIOps technologies continue to mature:

- Automated contextual data discovery
- Autonomous application of predictive models
- Automation of the prediction value chain delivery

6. Business Insights

So far, all the use cases have been mainly focusing on technology operations. However, we have seen an increasing trend in the use of AIOps to garner business insights. This is not surprising as operations sit at the center of every business, delivering values day by day. All the data collected can help make informed business decisions and forecast new trends which business can capitalize on. The focus on customers and their interaction with business are the most important criteria to succeed in today's competitive business landscape. Hence, it is expected that more investment will be made on AIOps to improve user experience.

There is no doubt that AIOps will continue to mature, disrupting IT operations management. The technology is used today to assure stability, reduce operational costs, and increase IT productivity. But increasingly, it has the potential to be the key differentiator for business to optimize their operations and maximize their ROI in Technology. As Technology will play a significant role in any business, this means AIOps will be at the core to any business operations. Hence, hang on tight and start investing in AIOps today!

Help Us to Improve This Book

Thanks for purchasing this book, we hope you have found value in it.

If you want to add value to this book and become a volunteer contributor, you can join our AIOps Book community. Visit https://classroom.google.com and join this classroom: **ibjk4ru**

We want to keep updated this book and publish new future editions.

Help us to make the book even better and change the world.

We will give credit to all contributors.

Bibliography and Credits

- CA. The Definitive Guide to AIOps, 2019
- Cisco. Service-Centric Approach to AIOps, 2019
- Cloudfabric. Going Beyond AIOps to Accelerate IT Transformation, 2019
- Gavstech. Algorithmic IT Operations (AIOps) Platforms for Unique Business Insights, 2019
- ITConcepts. AIOps–a technical overview, 2019
- Moogsoft. Everything You Need to Know About AIOps, 2019
- Red Hat. AIOps: Anomaly detection with Prometheus, 2019
- Savision. Digital Transformation   Powered by AIOps, 2019
- Baidu. Next Generation of DevOpsAIOpsin Practice, 2019
- FixStream. AIOps for Dummies, 2016
- Will Cappelli. Why AIOps must move from Monitoring to Observability, 2018

www.ingramcontent.com/pod-product-compliance
Lightning Source LLC
Chambersburg PA
CBHW021153020426
42331CB00003B/37